Quantification of Tannins in Tree and Shrub Folia‹

Quantification of Tannins in Tree and Shrub Foliage

A Laboratory Manual

by

Harinder P.S. Makkar

Animal Production & Health Section,
Joint FAO/IAEA Division of Nuclear Techniques in Food and Agriculture,
International Atomic Energy Agency, Vienna, Austria

KLUWER ACADEMIC PUBLISHERS
DORDRECHT / BOSTON / LONDON

A C.I.P. Catalogue record for this book is available from the Library of Congress.

ISBN 1-4020-1632-8

Published by Kluwer Academic Publishers,
P.O. Box 17, 3300 AA Dordrecht, The Netherlands.

Sold and distributed in North, Central and South America
by Kluwer Academic Publishers,
101 Philip Drive, Norwell, MA 02061, U.S.A.

In all other countries, sold and distributed
by Kluwer Academic Publishers,
P.O. Box 322, 3300 AH Dordrecht, The Netherlands.

Printed on acid-free paper

Printed in the Netherlands.

FOREWORD

Tanniniferous trees and shrubs are of importance in animal production because they can provide significant protein supplements, but unfortunately the amounts of tannins that they contain vary widely and largely unpredictably, and their effects on animals range from beneficial to toxicity and death. The toxic or antinutritional effects tend to occur in times of stress when a very large proportion of the diet is tanniniferous. With a better understanding of tannin properties and proper management, they could become an invaluable source of protein for strategic supplementation. As the demand for food rises, tanniniferous plants must play an increasingly important part in the diet of animals, in particular for ruminants in small-holder subsistence farming in developing countries. It is therefore critical that techniques be developed to measure and manage the anti-nutritional components that they contain.

Keeping the above in mind, a Joint FAO/IAEA Co-ordinated Research Project (CRP) on "Use of Nuclear and Related Techniques to Develop Simple Tannin Assays for Predicting and Improving the Safety and Efficiency of Feeding Ruminants on Tanniniferous Tree Foliage" was initiated in 1999. In order to provide a sound basis for this CRP, an FAO/IAEA Consultants Meeting was held in August 1997 in Vienna, at which the tanniniferous plants to be studied, the analytical methods, the test animals and the animal response evaluation techniques were defined.

This publication contains methodologies for the analysis of tannins using chemical-, protein precipitation/binding- and bio-assays recommended by the consultants. Some of these methods were used in the first phase of the CRP (1999–2001) with the aim of identifying assays that predict the biological response of feeding tannin-containing plants. A summary of the results achieved in the first phase of the CRP and the objectives of the on-going second phase of this project are presented at the end of this manual. It is hoped that this publication will contribute to the strategic and efficient utilization of tannin-containing tree and shrub foliage and agro-industrial by-products.

James D. Dargie
Director, Joint FAO/IAEA Division
of Nuclear Techniques in Food and Agriculture
Vienna, Austria

ACKNOWLEDGEMENTS

The contributions of Drs. Ann Hagerman, Ellen Hoffmann, and Irene Mueller-Harvey for the radio-labelled assay and the hydrolysable tannin assay using potassium iodate, the protein precipitation assay based on image analysis, and characterization of phenolic compounds using thin layer chromatography, respectively, are gratefully acknowledged.

EDITORIAL NOTE

The use of particular designations of countries or territories does not imply any judgement by the publisher, and the IAEA, as to the legal status of such countries or territories, of their authorities and institutions or of the delimitation of their boundaries.

The mention of names of specific companies or products (whether or not indicated as registered) does not imply any intention to infringe proprietary rights, nor should it be construed as an endorsement or recommendation on the part of the IAEA.

CONTENTS

Foreword i
Acknowledgements iii
Editorial Note v

1. CHEMICAL, PROTEIN PRECIPITATION AND BIOASSAYS FOR TANNINS, TANNIN LEVELS AND ACTIVITY IN UNCONVENTIONAL FEEDS, AND EFECTS AND FATE OF TANNINS ... 1

1.1. Introduction ... 2
1.2. Tannin assays ... 4
1.2.1. Chemical methods ... 4
1.2.1.1. Total phenols ... 4
1.2.1.2. Tannins ... 5
1.2.1.3. Condensed tannins ... 5
1.2.1.4. Hydrolysable tannins ... 7
1.2.2. Gravimetric methods ... 9
1.2.3. Protein precipitation methods ... 10
1.2.4. A tannin bioassay ... 11
1.2.5. ^{14}C-labelled polyethylene glycol binding assay ... 12
1.2.6. Near infrared based method ... 13
1.2.7. Which standard should be used for chemical and protein precipitation assays? ... 13
1.2.8. Screening of condensed and hydrolysable tannins and characterization of phenolic by thin layer chromatography ... 16
1.3. Level and activity of different types of tannins in feedstuffs ... 17
1.3.1. Leaves from trees and browses ... 17
1.3.2. Fallen oak leaves ... 20
1.3.3. Weeds ... 21
1.3.4. Legume straws and foliage ... 21
1.3.5. Faba beans ... 21
1.3.6. Agro-industrial and forestry by-products ... 24
1.4. Effects and fate of tannins in ruminants ... 25
1.4.1. Digestion kinetics ... 25
1.4.2. Antinutrient-antinutrient interactions ... 26
1.4.3. Biological significance of bound condensed tannins ... 27
1.4.4. Efficiency of microbial protein synthesis ... 27
1.4.5. Fate of tannins ... 29
1.4.6. Adaptation to tannins ... 31
1.4.7. Post-absorptive effects ... 32
1.5. References ... 34

2. TREATMENT OF PLANT MATERIAL, EXTRACTION OF TANNINS, AND AN OVERVIEW OF TANNIN ASSAYS PRESENTED IN THE MANUAL 43

2.1. Introduction 43
2.2. Preparation of plant extract 43
2.2.1. Collection, drying and storage of plant material 43
2.2.2. Grinding of sample 44
2.2.3. Extraction of tannins 44
2.2.4. Suggested procedures 45
2.3. Tannin assays 46
2.3.1. Chemical methods 46
2.3.2. Protein precipitation/binding methods 47
2.3.3. Tannin bioassay 48

3. MEASUREMENT OF TOTAL PHENOLICS AND TANNINS USING FOLIN-CIOCALTEU METHOD 49

3.1. Reagents 49
3.2. Preparation of calibration curve 50
3.3. Analysis of total phenols 50
3.4. Removal of tannin from the tannin-containing extract 50

4. DETERMINATION OF CONDENSED TANNINS (PROANTHOCYANIDINS) 53

4.1. Reagents 53
4.2. Analysis 53

5. GALLOTANNIN DETERMINATION BY RHODANINE ASSAY 55

5.1. Reagents 55
5.2. Determination of free gallic acid 55
5.3. Determination of gallic acid present in free and in gallotannin forms 55
5.4. Calibration curve 56

6. GALLOTANNIN DETERMINATION USING HPLC 57

6.1. Preparation of sample for determination of free gallic acid 57
6.2. Preparation of sample for determination of gallic acid present in free and in gallotannin forms 57

7. DETERMINATION OF HYDROLYSABLE TANNINS (GALLOTANNINS AND ELLAGITANNINS) AFTER REACTION WITH POTASSIUM IODATE 59

7.1. Reagents 59
7.2. Procedure 59
7.2.1. Methanolysis and pH adjustment. 59
7.2.2. Standard and controls 60
7.2.3. Reaction with potassium iodate 60

8. DETERMINATION OF PROTEIN-PRECIPITABLE PHENOLICS.........63

8.1. Reagents.........63
8.2. Formation of the tannin-protein complex.........63
8.3. Determination of tannins (phenolics) in tannin-protein complex.........64
8.4. Protein precipitable phenolics as percentage of total phenolics.........65
8.4.1. Determination of total phenolics in the original extract.........65
8.5. Determination of proteins in tannin-protein complex using image analysis.........66
8.5.1. Sample preparation.........66
8.5.2. Modified dot blot assay (amido black).........66
8.5.3. Calibration.........67
8.5.4. Procedure.........67
8.5.5. Membrane Holder.........68
8.5.6. Application pattern.........69
8.5.7. Stained membrane with regions defined for evaluation.........69

9. PROTEIN-BINDING CAPACITY BY FILTER PAPER ASSAY.........71

9.1. Reagents.........71
9.2. Preparation of plant extract.........71
9.3. Assay procedure.........71

10. RADIAL DIFFUSION ASSAY.........73

10.1. Chemicals.........73
10.2. Reagents.........73
10.3. Preparation of plates.........73
10.4. Assay procedure.........73

11. TANNIN BIOASSAY.........75

11.1. Sample preparation.........75
11.2. Reagents.........75
11.3. Weighing of samples and preparation of syringes.........75
11.4. Preparation of *in vitro* rumen fermentation buffer solution.........76
11.5. Measurement of microbial protein in the residue (apparent undigested residue) left after fermentation.........77
11.5.1. Preparation of apparent undigested residue.........77
11.5.2. Preparation of lyophilized rumen microbial fraction.........78
11.5.3. Determination of purines (marker for microbial mass/protein).........78

12. RADIOLABELLED BSA PRECIPITATION METHOD.........83

12.1. Reagents.........83
12.2. Supplies.........84
12.3. Procedure for BSA precipitation (amount of BSA precipitated by tannic acid or plant extracts).........85
12.4. Calculations.........85

13. SIMPLIFIED RADIOLABELLED BSA PRECIPITATION METHOD....87

13.1. Reagents....87
13.2. Procedure for protein binding....87
13.2.1. Immobilization of tannins from the plant extract....87
13.2.2. Protein binding on immobilized tannins....87

14. CHARACTERIZATION OF PHENOLIC COMPOUNDS BY THIN LAYER CHROMATOGRAPHY (TLC)....89

14.1. Characterization of condensed tannins....89
14.1.1. Samples....89
14.1.2. Reagents....89
14.1.3. Material for TLC....89
14.1.4. Extraction....90
14.1.5. Procedure....90
14.2. Survey of condensed and hydrolysable tannins by TLC....91
14.2.1. Solvents....91
14.2.2. Spray reagents to detect different classes of tannins....91
14.3. TLC procedure....93

15. REFERENCES....95

16. A SUMMARY OF RESULTS FROM THE FIRST PHASE OF THE CO-ORDINATED RESEARCH PROJECT....97

16.1. The tannin assays and standards used in the first phase....97
16.2. Conclusions....98
16.3. Recommendations....99
16.4. Objectives and studies for the on-going second phase....99

17. LIST OF PARTICIPANTS IN THE CO-ORDINATED RESEARCH PROJECT....101

1. CHEMICAL, PROTEIN PRECIPITATION AND BIOASSAYS FOR TANNINS, TANNIN LEVELS AND ACTIVITY IN UNCONVENTIONAL FEEDS, AND EFECTS AND FATE OF TANNINS

Abstract

CHEMICAL, PROTEIN PRECIPITATION AND BIOASSAYS FOR TANNINS, TANNINS IN UNCONVENTIONAL FEEDS, AND EFFECTS AND FATE OF TANNINS.

Animal production systems in the tropical and subtropical countries utilize a wide range of feedstuffs; the main amongst these being the crop and industrial by-products, grasses, legumes, trees and shrubs. Trees and shrubs are of importance in animal production because they can provide significant protein supplements, especially in the dry season. But, the amount of tannins that they contain vary widely and largely unpredictably, and their effects on animals range from beneficial to toxicity and death. Similarly many agro-industrial by-products contain tannins. With a better understanding of tannin properties and proper management, they could become invaluable source of protein for strategic supplementation. As the demand for food rises, these unconventional feedstuffs must play an increasingly important part in the diet of animals, in particular for ruminants in small-holder farming in developing countries. It is therefore critical that proper techniques be used to measure and manage the anti-nutritional effects they cause. The analysis of tannins remains highly problematic. Various chemical assays for hydrolysable tannins and condensed tannins are available. Most tannin-containing feedstuffs contain both hydrolysable tannins and condensed tannins, but unfortunately only condensed tannins are generally analysed, probably because of simplicity of the assays measuring these; and the feedstuffs termed as tannin-containing or tannin-free feedstuffs. Furthermore, the biological effects are ascribed to mostly the condensed tannins (or the synonym: proanthocyanidins), which might be misleading. This paper reviews the available assays for hydrolysable tannins and condensed tannins and highlights the advantages and disadvantages of each. Protein precipitation assays, both isotopic and non-isotopic, representing the operational property of both hydrolysable tannins and condensed tannins; a tannin bioassay based on an *in vitro* simulation of the rumen and measurement of tannin activity for both free and bound tannins in terms of rumen fermentation parameters; and ^{14}C-polyethylene glycol binding assay are also discussed. Each type of tannin responds differently in each of these assays. This variability makes it impossible to use any single method. Use of a battery of methods, therefore, is suggested; and these assays are being used in the FAO/IAEA-sponsored projects on the utilisation of tree foliage as livestock feed. Using these assays for 37 shrub and tree leaves, highly significant correlation existed between protein precipitation capacity and extractable total phenols (r = 0.87) or tannins (r = 0.83). On the other hand, a weak correlation was observed between condensed tannins (measured by the butanol-HCl method) and protein precipitation capacity (r = 0.41), which could be due to the variation in structural and biological activity of tannins. The correlations observed between extractable total phenols, tannins or condensed tannins and the tannin bioassay values based on the rumen simulation technique were similar to those obtained between extractable total phenols, tannins or condensed tannins and protein precipitation capacity. Highly significant correlations between extractable phenolics or tannins with protein precipitation capacity or the values obtained using the tannin bioassay suggest that extractable total phenolics and tannins values could be taken as a measure of biological activity of tannins. The condensed tannins values by the butanol-HCl-iron method do not appear to reflect the biological activity. From the relationships between chemical, protein precipitation and bioassays, it was postulated that tree and shrub leaves with extractable total phenol and tannin contents of approximately 4.5% and 2.0% respectively (as tannic acid equivalent) will not produce significant adverse effects on ruminant livestock. The hydrolysable tannins, measured by an HPLC and a spectrophotometric method (rhodanine), were present in all the 37 samples (one having *ca* 14%; three between 1.5 and 3.5%; six *ca* 0.5% and the rest below 0.5%) analysed; and the hydrolysable tannins bind proteins, affect rumen fermentation, and could cause adverse effects similar to condensed tannins. Hydrolysable tannins are also known to be toxic and can cause death of animals if consumed in large quantities. The roles of rumen microbes in degradation and overcoming the detrimental effects of tannins, and the effects and fate of tannins in ruminants are also discussed. Simple methods based on postharvest technology, treatment with low-cost chemicals, biological treatments, and supplementation with tannin-complexing agents, to enhance the feeding value of tannin-containing feeds are needed.

1.1. Introduction

In economically developed regions, livestock products, such as meat, milk, eggs and hides, account for more than one-half of the value of total agricultural production. In most developing regions the proportional value of livestock products is lower but still appreciable. As a proportion of total agricultural production, livestock products amount to about 22% for Southeast Asia, 25% for sub-Saharan Africa (not including the Republic of South Africa), 26% for China, 31% for West Asia and North Africa and 38% for South America. The primary constraint to livestock production in developing countries is the scarcity and fluctuating quantity and quality of the year-round feed supply. These countries experience serious shortages of animal feeds and fodders of the conventional type. Natural forages are very variable both in quality and quantity, conventional agro-industrial by-products are scarce and vary seasonal and grains are required almost exclusively for human consumption.

Multipurpose trees and non-conventional agro-industrial and forestry by-products can play a vital role in bridging the wide gap that exists between supply and demand for feeds. Feed production options such as forage legumes and multi-purpose trees are often environmentally beneficial. Their use by livestock brings economic benefits to farmers as well as improvements in soil fertility and control of soil erosion. In addition neither multipurpose trees nor non-conventional agro-industrial by-products can be considered as food for humans. The proper use of these feedstuffs is hampered due to the presence of factors, which are loosely addressed as antinutritional factors, plant defensive components or plant secondary substances. These compounds do not function in primary metabolism such as energy conversions, biosynthesis or biodegradation, but appear to have a diverse roles ranging from protecting plants from diseases and herbivore attack to toxicity and mimicking hormone actions. These components include phenolics, saponins, alkaloids, free amino acids, steroids, essential oils, glycosides, terpenes, and resins. The most widely occurring components from these groups are tannins Tannins are polyphenolic substances with various molecular weights and a variable complexity. These are chemically not well defined substances but rather a group of substances with the ability to bind proteins in aqueous solution. Their multiple phenolic hydroxyl groups lead to the formation of complexes primarily with proteins and to a lesser extent with metal ions, amino acids and polysaccharides. Tannins are tentatively classified into two classes: hydrolysable and condensed tannins (although tannins are known which have components of both hydrolysable and condensed tannins), and are considered to have both adverse and beneficial effects depending on their concentration and nature besides other factors such as animal species, physiological state of the animal and composition of the diet. Although research on tannins has a long history, considerable additional research must be carried out in order to fully exploit benefits of incorporating tree leaves and agro-industrial by-products in livestock feed. It is imperative to generate information on the content and nature of tannins present in these feedstuffs, and on effects and fate of tannins in animals. As tannins are considered to play an important role in the plant's defence against environmental stresses and in disease resistance, the tree leaves and agro- and forestry-based by-products in the tropics are likely to have high levels of tannins. High levels of tannins are known to produce adverse effects. Another important challenge has been to develop simple and economically viable detanninification methods for these tannin-containing unconventional feed resources. This article contains a synthesis of the work conducted in the areas related to the Joint FAO/IAEA Coordinated Research Project on tannins under which this manual is being published. The areas covered are: a) tannin assays, b) tannin level and activity in some unconventional feeds, and c) effects and fate of tannins in animals. The latter work has generated information on: i) the effects of tannins on rumen fermentation in particular enzyme activities, digestion kinetics of feedstuffs, availability and partitioning of nutrients, efficiency of microbial protein synthesis, ii) biochemical and toxicological parameters associated with oak poisoning, iii) nutritional significance of bound condensed tannins, iv) applicability of detergent system of fibre analysis for tannin-rich feed samples, v) degradation of condensed tannins by rumen microbes, vi) physiological significance of salivary proline-rich proteins in the adaptation of cattle to tannins, and vii) anti-carcinogenic activity of tannins isolated from tree leaves.

1.2. Tannin assays

Use of proper analytical techniques for measurement of tannins is vital in order to meet the above challenges. A considerable number of different assays have been developed for the measurement of tannins. However, these assays, due to the complex and diverse nature of tannins, do not provide satisfactory results. Each method measures different types of tannins based on chemistry of the reaction between tannins/phenols and reagents used. The methods generally in use are categorised into two groups: chemical methods and protein precipitation methods. In addition to these two conventional categories of tannin assays, other assays such as gravimetric assays, a tannin bioassay based on the *in vitro* gas method and inclusion of polyethylene glycol (a tannin-binding agent), and ^{14}C labelled polyethylene glycol binding assay have been discussed.

1.2.1. Chemical methods

The most commonly used procedures in this category are: the redox methods (Folin-Ciocalteu, Folin-Denis or Prussian blue methods), the vanillin assay, the metal complexing assay, and the acid butanol assay with and without addition of iron [1].

1.2.1.1. Total phenols

The Folin-Denis, Folin-Ciocalteu or Prussian blue methods are used to measure total phenols. These assays are based on oxidation of phenolic analyte and reduction of the reagent to form a chromophore. Presence of reducing agents such as ascorbic acid, amino acids, xanthine, proteins etc. interfere in the assay. These methods provide neither a means to distinguish tannins (phenolics which precipitate proteins) from nontannin phenolics, nor a means to identify specific types of tannins in a mixture (Fig.1). Similarly, the metal complexing method (commonly known as ferric chloride assay) based on formation of coloured phenolic-metal ion complexes is also useful for measuring total phenols. Interference from nonphenolics is unlikely, however, nontannin phenolics cannot be distinguished from tannins with this method. Using the Folin-Denis, Folin-Ciocalteu or Prussian blue methods, the results are generally expressed as tannic or gallic acid equivalent. Amongst these methods, we found the Folin-Ciocalteu method to be highly reproducible. This method is also the most sensitive and the Folin-Ciocalteu reagent is commercially available.

1.2.1.2. Tannins

In the method [2], tannins are distinguished from nontannins by the use of a solid matrix, polyvinylpolypyrrolidone (PVPP). This method assumes that the phenolics, which bind to proteins, are the same as those which bind to PVPP. Total phenols are measured in a plant extract using the Folin-Ciocalteu method before and after treatment with PVPP. PVPP has a high affinity for tannins and therefore removal of PVPP, following the treatment, using centrifugation removes tannins from the extract. The difference between total phenol values before and after the PVPP treatment is a measure of tannins. The ferric chloride assay could also have been used in place of the Folin-Ciocalteu assay for measurement of total phenols before and after the PVPP treatment. The former is preferred because of its higher sensitivity (range of calibration curves: 2–10 μg vs 25–125 μg tannic acid for the ferric chloride assay; see page 296 of [2]). The above method based on PVPP treatment [2] is capable of measuring total tannins. It cannot identify specific types of tannins in a mixture, i.e. information on presence or absence of condensed or hydrolysable tannins cannot be obtained by this method.

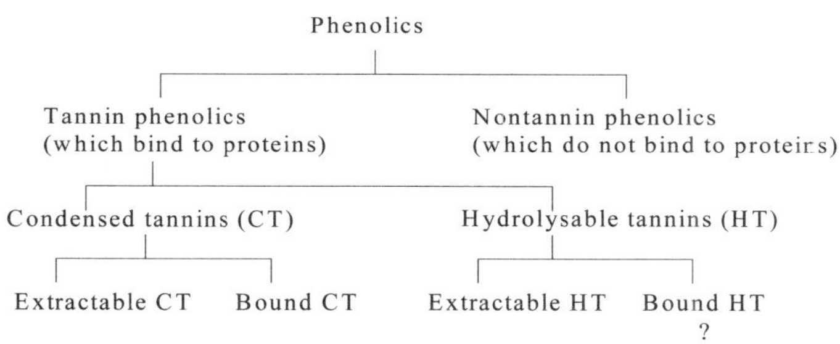

FIG.1. Categorization of plant phenolics

1.2.1.3. Condensed tannins

Condensed tannins (*synonym*: proanthocyanidins; hydrolytic cleavage of condensed tannins yields anthocyanidins) comprise a group of polyhydroxyflavan-3-ol oligomers and polymers linked by carbon-carbon bonds between flavanol subunits. The vanillin assay in methanol [3, 4] has been widely used for measuring condensed tannins in sorghum and other beans. This assay measures not only condensed tannins but also flavan-3-ols and dihydrochalcones which are nontannin. Other disadvantages of this method are that monomeric flavans give higher colour yield as compared to condensed tannins and proanthocyanidins based on 5-deoxyflavanols, such as profistinidin in quebracho tannin do not react. We found that presence of acetone interferes in this assay. Acetone was found to form chromogen with acidified vanillin with an absorption peak at 548 nm, which produces a substantial error in the determination of condensed tannins. In addition, reproducibility of this method is not good [5, 6]. Because of these problems associated with this method, it is not recommended for tree leaves and agro-industrial by-products, as extractability of tannins from these feedstuffs is higher in 70% aqueous acetone [7–9]. For sorghum and other beans for which this method has been widely used, the extracts were prepared in methanol; methanol does not interfere in this method [6].

The butanol-HCl method originally proposed [10] for measurement of condensed tannins is simple and more specific compared to the vanillin assay. Later, this method was modified by inclusion of iron in the butanol-HCl reagent and showed higher sensitivity and better reproducibility of the assay in presence of iron [11]. This method is based on oxidative cleavage of the interflavan bonds in presence of mineral acids in alcoholic solutions at about 95°C to yield pink-coloured anthocyanidins, which are measured at 550 nm. This method is sensitive to the presence of water; increase in the water content decreases the colour yield of anthocyanidins. Tannins besides being extractable also exist in the bound form, as contaminants in the fibre or protein fractions. The bound proanthocyanidins or condensed tannins are generally measured by the butanol-HCl-iron reagent. The hydrolysis of bound proanthocyanidins to anthocyanidins is a prerequisite for determination of these condensed tannins. For accurate determination of these bound tannins, the samples should be freeze-dried because drying, even at temperatures as low as 50°C, could decrease the values obtained [12]. Heat treatment of tannins favours chemical bond formation between tannins and other macromolecules e.g., phenolic groups could get oxidised, leading to quinones which, in turn, can give rise to condensation reactions with other macromolecules. These altered 'tannins' no longer release anthocyanidins or release anthocyanidins to a lesser extent leading to underestimation of bound proanthocyanidins/condensed tannins. Even in freeze-dried samples, not all bound condensed tannins react quantitatively in the butanol-HCl reagent, leading to underestimation of bound condensed tannins [13]. The butanol-HCl method should be used with caution as a quantitative assay. The values obtained using the method do not seem to correlate with the biological value of tannin-containing feeds (see below). This method, nevertheless, is a simple method to know the presence of condensed tannins in feedstuffs. Terrill *et al.* [14] also proposed a method, based on the butanol-HCl reagent, for determination of extractable, protein- and fibre-bound condensed tannins. The classification of condensed tannins as bound to protein and fibre by this method has not been validated (protein and fibre bound condensed tannins as determined by this method are really the condensed tannins bound to fibre and protein) and could be arbitrary and loosely categorised. In addition the nutritional significance of the values obtained for fibre and protein bound has not been evaluated. The presence of hydrolysable tannins in the bound form in feedstuffs has not yet been shown, although these are present in the heartwood of *Castanea sativa* and *Quercus petraea* [15].

Other methods for the analysis of insoluble tannins include the use of ^{13}C-NMR [13], thiolysis and phloroglucinol degradation [16]. These methods because of their complexity have not been used much in the past.

The degree of polymerization of proanthocyanidins can be measured from the ratio of the results from the vanillin assay in glacial acetic acid (in which only terminal units react to form chromophore) and those from the butanol-HCl assay (in which only extender units react to form chromophore) [17]. This method can only be used to compare chemically similar tannins, since reactivity in the butanol-HCl assay is the function of the interflavan bond. This method, therefore, measures relative degree of polymerisation, and has been used to investigate nature of condensed tannins in isolated tannins [18] and to study changes in the nature of condensed tannins as affected by maturity of leaves and detoxification processes [7–9, 19–23]. A simple method put forward for characterization of proanthocyanidins and determination of degree of their polymerisation based on acidic degradation of proanthocyanidins with phenylmethanethiol (thiolysis) followed by HPLC [24,25] could offer valuable information on structure-activity relationship of proanthocyanidins in feeds.

1.2.1.4. Hydrolysable tannins

Hydrolysable tannins are esters of two phenolic acids, gallic acid and hexahydroxydiphenic acids. An approach employed in some methods for determination of hydrolysable tannins is the conversion of the hydrolysable tannins to a common unit and then determination of the common units by spectrophotometric or HPLC methods. This is similar to the degradation of flavonoid subunits of condensed tannins to anthocyanidins that is determined spectrophotometrically by the butanol-HCl method (see above). These methods are useful for simple hydrolysable tannins but may provide limited information for complex oligomeric hydrolysable tannins.

The rhodanine method [26] determines gallotannins as gallic acid equivalent. In this method gallotannins are acid hydrolysed under anaerobic conditions to gallic acid which is reacted with rhodanine to give pink chromophore (measured at 520 nm). The free gallic acid (before the acid hydrolysis) is also measured, which is subtracted from the total gallic acid (gallotannins plus free gallic acid) measured after the hydrolysis to obtain gallotannins as gallic acid equivalent. The number of gallic acid units differs between different gallotannins, and therefore this method does not provide an absolute quantification of gallotannins. In addition, some ellagitannins are also known to contain gallic acid. The specificity and sensitivity of the assays could be increased by measuring gallic acid before and after the acid hydrolysis by the HPLC method. The details of this method [27] are available in this manual. It uses Nucleosil 120-5 C18 column, and the separation is made at room temperature (ca 22°C) at a flow rate of 1.2 ml/min using a gradient elution. Gallic acid is eluted between 14 and 15.5 min and detected at 280 nm.

The rhodanine and HPLC methods have been used for determination of gallotannins in a large number of samples (see below).

The sodium nitrite method [28] for determination of ellagitannins, based on determining the product of hydrolysis of hexahydroxydiphenic acid esters (ellagitannins) requires large quantities of pyridine as a solvent, introducing significant toxicity risk when the method is used for routine analyses of large sample sets.

Postassium iodate method [28] measures both gallotannins and ellagitannins and is used occasionally but neither the reaction chemistry nor specificity of the method is well established. The common structural moiety for all of these compounds is gallic acid, which can be easily produced by acid hydrolysis of the compounds. However, only gallate esters react to form the chromophore which is monitored in the potassium iodate method. Gallic acid is oxidized by potassium iodate to form a yellow compound that has similar spectral properties to the brownish products formed by flavonoids and other phenolics upon extensive oxidation, and is thus not useful for selective determination. In addition, a transient reaction product with maximum absorbance between 500–550 nm is obtained when galloyl esters were reacted with potassium iodate in the presence of either acetone or methanol. Several practical limitations of the method include the inconvenient temperature for the reaction (requires a refrigerated water bath); the necessity of running each reaction for a different period of time to achieve maximal colour yield; the occurrence of interfering yellow oxidation products; formation of several chromophores with different lambda maxes for some plant extracts comprised of mixtures of tannins; and the formation of precipitates when extracts from some plants are analyzed. Despite these limitations, this method has been used occasionally for species rich in hydrolysable tannins, to provide an estimate of these compounds in crude plant extracts. Recently, Hagerman and her co-workers have modified the potassium iodate method to include a first step in which all of the hydrolysable tannins in the sample are converted to a single chemical species, methyl gallate [29]. Differences in reactivity of the parent compounds are thus overcome. Results from various laboratories can be compared directly since they are reported in terms of a commercially available standard compound, methyl gallate. In the modified method the hydrolysis conditions are changed to use methanol as the solvent rather than water, which yields methyl gallate rather than gallic acid (aqueous medium yields gallic acid). Like the more complex galloyl esters comprising the hydrolysable tannins, methyl gallate reacts with potassium iodate to form a red chromophore. The conditions for methanolysis have been established to assess the temperature and time required to maximize yield of methyl gallate. The temperature for methanolysis has been fixed at 85°C to avoid disappearance of some methyl gallate and formation of unidentified phenolic products observed at higher temperatures especially in the crude plant extracts. Under the methanolysis conditions, methyl gallate was stable for at least 24 h. For routine analysis, 20 h has been suggested as a convenient time for methanolysis. Presence of water during the methanolysis creates problem in the method — conversion to methyl gallate is not quantitative and some gallic acid is formed. Other precaution is that the caps should be tightly fit on the tube during methanolysis in order to avoid evaporation of the methanol. In addition, the reaction conditions between methyl gallate and potassium iodate have been optimized to form stable chromogen. These modifications have improved substantially the utility of the method for measuring hydrolysable tannins ([29]; details of the method are presented in this manual). A similar approach has been used [30] in which anhydrous methanolic HCl has been used for releasing ellagic acid and gallic acid (as methyl gallate) followed by determination of individual moieties using HPLC. This method could also be adapted for measuring soluble and insoluble hydrolysable tannins in feed resources.

1.2.2. Gravimetric methods

None of the above methods determines tannins in absolute terms but measure their concentrations relative to one or another standard, namely tannic acid or gallic acid in methods based on the oxidation-reduction principle; and catechin, quebracho tannins and leucoanthocyanins in methods for the determination of condensed tannins. A method for quantifying tannins gravimetrically [2] was devised in order to overcome the possible overestimation of dry matter digestibility in tannin-rich feeds when determined gravimetrically as in the nylon bag or in the Tilley and Terry method. Under such experimental conditions tannins are solubilized but might still be indigestible or partially digestible and, more importantly from a nutritional point of view, might not contribute to energy production when consumed by an animal. In order to correct for dry matter digestibility of tannin-rich feeds it is therefore imperative to have a method for quantification of tannins that measures tannins in absolute terms (by mass) and not relative to a standard. The method [2] is based on weighing the tannin extract before and after removal of the tannin by treatment with insoluble polyvinylpyrrolidone (PVPP) to bind tannin and removal of the PVPP-tannin complexes by centrifugation. The values obtained for tannins are used for determination of the 'true' dry matter digestibility of tannin-rich feeds. Gravimetric methods do not suffer from some of the disadvantages associated with colorimetric methods. In particular, they do not rely on comparison with a standard tannin which may be quite different from the tannins present in the plant of interest. The author would like to respond to a comment on this method made recently [31] that '30 mg of PVVP has been used to bind 2 mg of tannin acid completely, and the percentage weight change in insoluble PVPP is thus quite small'. It is not the change in the PVPP weight that is measured in the gravimetric method [2] but the change in the weight of the extract before and after the PVPP treatment, and that too using a large amount of the extract and the PVPP. In the spectrophotometric method based on this principle (see above) 30 mg PVPP is used to bind tannins of the order of 2 mg and then the total phenols before and after the PVPP treatment are measured in the extract using the Folin-Ciocalteu method.

Gravimetric methods are not necessarily perfect. They have generally lower sensitivities than colorimetric methods and are time consuming. The gravimetric method [2] also suffers from these disadvantages and so its use is recommended when the dry matter digestibility of tannin-rich feeds needs to be corrected. For other routine applications, the principle of the gravimetric method is used in conjunction with a spectrophotometric method. Total phenols are determined spectrophotometrically using the Folin-Ciocalteu reagent in the extract before and after the PVPP treatment. The difference in these phenolic values is a measure of tannins [2], and this difference (tannins) when expressed as a tannic acid equivalent (Merck, Darmstadt, Germany) was quite close to the tannin levels determined gravimetrically in leaves from various trees and browses; each gram of tannins (by mass) had a reducing power equivalent to 0.76 to 1.25 g tannic acid [2].

The gravimetric method [32] based on the precipitation of phenolics by ytterbium acetate measures total phenolics and not tannins. In addition this method is not specific for phenolics; it precipitates other moieties in addition to phenolics, and the precipitation is not complete at low phenolic concentration and for some phenolics, for example, rutin is not precipitated [33] leading to underestimation of phenolics.

1.2.3. Protein precipitation methods

Tannins have the ability to precipitate proteins and based on this property of tannins many methods for their determination have been developed [34]. The methods for quantification of tannins based on their operational properties i.e., their capacity for complexing with proteins (protein precipitation assays) are considered to provide better information on the biological value of feeds and fodders containing tannins. The advantages and disadvantages of the methods that are based on this principle e.g., enzyme inhibition and those that measure protein precipitable phenolics and not the protein precipitation capacity of tannins by measuring protein in the tannin-protein complex have been discussed in [34]. The method [35] based on binding of ^{125}I-labelled bovine serum albumin for determination of the protein precipitation capacity of tannins is accurate and sensitive. However, this method requires special equipment and some degree of expertise, because the method is a radio-assay. In a protein precipitation method for the determination of tannins [36], the protein in the tannin-protein complexes has been measured using the ninhydrin assay of amino acids released by alkaline hydrolysis of the complex. Subsequently, this method has been modified to enable measurement of both protein and phenolics in a tannin-protein complex [37]. The protein was measured by the ninhydrin reaction as mentioned above [36] and the phenolics by the ferric chloride method [38]. This enabled measurement of the ratio of protein to tannin in the tannin-protein complex, which can be defined as the specific activity of tannins. This represents the amount of protein bound by a unit of phenolics/tannins in the complex, which could provide valuable information regarding changes in the nature of tannins, vis-a-vis their protein binding capacity associated with, for example, development, maturity, postharvest storage and detanninification of foods and feeds. In addition, we have found the method [37] to be much easier compared to method [36]. These two methods or as a matter of fact other protein precipitation methods measure tannins but do not allow selective determination of various types of tannins. In addition these methods are unable to determine the protein binding capacity if the concentration of tannins is low. In the method [39], tannins, phenols or a plant extract containing these moieties are applied on a chromatography paper and reacted with the protein bovine serum albumin. Unbound protein is washed off, and the protein in the tannin-protein complex is stained with Ponceau S, a dye specific for proteins. Protein-bound dye is eluted and the optical absorbance of the eluate is measured at 525 nm and converted to protein by the use of a calibration curve. This method also has the advantage, that the protein binding capacity of both hydrolysable and condensed tannins in an extract can be measured. This was the first method that enabled measurement of the contribution of the two types of tannin in an extract towards the total protein binding capacity of a plant extract/mixture of tannins. This method has provided valuable information on the role of hydrolysable tannins in oak poisoning (see below). The method [18] can also measure the protein binding capacity of hydrolysable and condensed tannins in a plant extract. However, it is much less sensitive than the method [39]. The sensitivity of the chromatography paper-protein-Ponceau S method [39] is also about 20 times higher (protein binding capacity of tannic acid as low as 5 μg can be determined) than that of the previous two methods [36, 37]. A further advantage is that, acetone containing plant extracts can be used in the chromatography paper-protein-Ponceau S method [39] whereas the presence of acetone interferes in other protein precipitation assays [36, 37]. The preparation of plant extract in aqueous acetone (generally 70%) is desirable because the solubility of tannins and phenolics is higher in aqueous acetone solution [7, 8], and acetone prevents oxidation of phenols.

Another protein precipitation method, which is insensitive to acetone, is the radial diffusion assay [40], in which tannin molecules migrate through agarose gel which is impregnated with the protein, bovine serum albumin (BSA). The tannin-protein complex is formed in the gel, which appears as a ring. The diameter of the ring is a measure of protein precipitation/binding capacity of tannins.

In most protein-binding assays the protein used is BSA, because it has the advantage of being inexpensive, easily available and soluble. However, these methods can be used with any other protein.

1.2.4. A tannin bioassay

All available protein precipitation assays measure tannins under conditions (temperature, pH, ionic strength, nature of protein) different from those of the rumen, and therefore the results obtained have limited applicability for predicting the nutritive value of tannin-containing feedstuffs. Polyvinylpyrrolidone and polyethylene glycol bind to tannins and make them inert [41]. This property has been exploited together with the *in vitro* gas method to quantify the effects of tannins on rumen fermentation [42]. Amongst various tannin-complexing agents, polyethylene glycol (MW 6,000) was found to be the most effective in binding to tannins [41]. Incubation of polyethylene glycol 6,000 along with a tannin-containing feedstuff in the *in vitro* system increased gas production. The percentage increase in gas production represents the effects of tannins; the higher the percentage increase in gas production, the greater the effect. Two factors which led to higher gas production in the presence of polyethylene glycol 6,000 were identified as the tannin activity of the extractable tannins measured as protein precipitation capacity and the protein content of the feedstuff (proteins mimic the effect of polyethylene glycol 6,000). However, these two factors explained only 42% of the variation [42]. Other factors, which could also influence the increase in gas production, might be the content and activity of bound tannins. Inclusion of factors such as the true protein content of the feed (non-protein nitrogen and fibre bound protein do not bind tannins), the protein precipitation obtained with protein from the respective feeds and measures of the activity of bound tannins is likely to increase the extent to which gas production can be predicted.

The possibility of quantifying tannin effects in terms of metabolizable energy (ME) and organic matter digestibility (OMD) of a feed was also put forward [42]. The increase in the gas production on the addition of polyethylene glycol 6,000 was translated into ME and OMD using the equations from [43] assuming that these equations are valid for tannin-rich feedstuffs. Factors which could render this assumption invalid are: i) tannins may interact with the gut and modify the metabolism and digestive, absorptive, immune and endocrine functions, ii) tannins may affect the partitioning of nutrients, and iii) tannins may protect feed protein from degradation in rumen. Later studies [44] showed that measurement of the gas production alone could be misleading as tannins affect partitioning of nutrients. The measurement of microbial mass production should be combined with gas measurement to give a better insight into the effects of tannins. This is achieved using purine and ^{15}N measurements. Full potential of this bioassay can be exploited if both the gas production and microbial mass/protein production are measured. This principle of measuring both the gas and microbial mass production could be used for evaluation of not only tannins but also of potential rumen modulators, of supplementation strategies, and of varieties and cultivars of good nutritional value in breeding programmes.

1.2.5. ^{14}C-labelled polyethylene glycol binding assay

Polyethylene glycols (PEG) of molecular weights 4,000 or 6,000 have a very high affinity for both hydrolysable and condensed tannins over a wide range of pH [41]. In this method the feed samples (without extraction of tannins) are kept in contact with PEG spiked with ^{14}C-labelled PEG, and the radioactivity bound to feed sample is a measure of tannins; higher the activity, higher is the tannins. The method is reportedly useful since there is no need to extract tannins, and it is considered to be a measure of both bound and extractable tannins [45]. High correlations between the PEG-binding assay and protein precipitation capacity method [2, 37] have been obtained and the values obtained have been found to be useful in predicting negative effects of tannins on ruminal degradation [45]. A major source of error in this method could arise from the formation of soluble tannin-PEG complexes, which will not be recovered in the feed sample and hence will underestimate the tannin values. There is a need to study the extent of formation of soluble versus precipitable PEG-tannin complexes under the conditions of the assay (and to study the nutritional significance of these soluble complexes), and to standardise the assay for parameters such as optimum particle size of the sample, temperature and treatment time. Additionally, correlations should be established between the PEG-binding assay and protein precipitation assays or the tannin bioassay using the *in vitro* gas method or the increase in the nitrogen degradability of feeds on addition of PEG (see next paragraph) for a large number of samples. Inclusion of *in vivo* parameters such as intake, nitrogen balance and degradability, production parameters (growth, wool production, etc.) for these feeds in the correlation studies would reveal the usefulness of the *in vitro* methods in predicting nutritional and physiological effects of feeding tannin-containing diets to ruminants.

Increase in nitrogen degradability of a feed when incubated in an *in vitro* rumen fermentation system in presence of PEG is also a measure of tannin activity [46, 47] and this increase in nitrogen degradability has also been found to predict the effects of tannins in ruminants [47].

1.2.6. Near infrared based method

Tannins in legume forages were quantified using near infrared reflectance spectroscopy. The wavelength, 2.150 μm was found to be an indispensable wavelength for prediction of condensed tannins [48]. This method may be applicable to the determination of tannins in large sample sets of homogenous feeds such as forage legumes.

1.2.7. Which standard should be used for chemical and protein precipitation assays?

For tannin determination, two types of standards are generally used: i) absolute standards, i.e., purified tannins from the plant of interest, and ii) relative standards, generally commercially available tannins. The use of absolute standards besides being inaccurate due to their poorly defined nature (presence of hydrolysable tannins when condensed tannins are to be determined and vice-a-versa; see below) also suffers from other disadvantages, such as the facts that they are time consuming and not practicable for large numbers of samples. Tannic acid is the commonly used standard in assays based on the principle of oxidation-reduction, metal complexing and protein precipitation due to its easy commercial availability and low cost. Tannic acid being a hydrolysable tannin (gallotannin) cannot be used for methods such as the vanillin-HCl and butanol-HCl-iron methods used for the quantification of condensed tannins. Tannic acids from various commercial sources behave differently towards a number of tannin assays [49] showing that: i) tannic acid preparations differed from source to source, and ii) the use of tannic acids from different sources as standards resulted in different tannin levels for a given sample. It is suggested therefore that: i) the source of tannic acid used as a standard should be stated, ii) the assay values obtained by different workers should be compared with caution, and iii) wherever possible (e.g., in protein precipitation assays) the results should not be expressed as tannic acid equivalents. In protein precipitation assays the results can be expressed as the amount of protein precipitation per unit of tannins or per unit of feed/food sample. Various problems associated with the determination of tannins and their possible solutions are reviewed in [5].

For condensed tannins the most widely used relative standards are quebracho, cyanidin and delphinidin. Quebracho tannins must be purified using Sephadex LH-20 from the spray dried quebracho powder available commercially (used for tanning leather). We have found a large batch to batch variation in the preparation of 'purified' quebracho tannins using the Sephadex LH-20 method, making the preparation a poor standard for condensed tannin determination. Furthermore, quebracho tannins give a much lower colour yield compared to delphinidin and cyanidin; the colour yield of delphinidin and cyanidin also differs considerably. Different condensed tannins contain different extent of cyanidin and delphinidin and the use of delphidin or cyanidin as a standard will under or overestimate the condensed tannin content depending on the proportion of cyanidin and delphinidin in the condensed tannins. A better option appears to be to express the results as absorption units. It has been observed that most animal nutritionist do not feel comfortable with absorption units and hence we convert the absorption at 550 nm to leucocyanidin equivalent using its extinction coefficient. Using this approach the results are then expressed as percent leucocyanidin equivalent but the conclusions from the results are the same as those obtained from the absorption units since the conversion of absorption units to leucocyanidin equivalent is nothing but multiplication by a constant factor.

One advantage of having to isolate tannins [18] is that they can be used for other purposes such as studying the nutritional, physiological and ecological implications of tannins in biological systems. For example, tannins thus isolated from various tree leaves were used in our studies on their anticarcinogenic properties [50].

Tannins were isolated from the leaves of some trees and shrubs and their properties studied [18]. Isolated tannins were found to be similar with respect to protein precipitation capacity to those present in the plant extract, indicating that the biological activity of isolated tannins is similar to those present in the plant. These tannin preparations were a mixture of condensed and hydrolysable tannins and should not be used as an absolute standard for determination of condensed tannins. Similarly, the extinction coefficient obtained for such preparations of condensed tannins are not of any use in the determination of condensed tannins in the plants, as these would overestimate the condensed tannins. Despite these reservations, this approach has, unfortunately, been used by some workers (see [18]) for determining condensed tannins in various tree leaves without testing for the presence of hydrolysable tannins in the isolated preparations, which were used as a standard.

None of the above mentioned methods provide unambiguous information on the composition or size of tannins. Although NMR and FAB mass spectroscopy are powerful tools for analysis of proanthocyanidins and for their structure elucidation, it is not currently possible to obtain useful mass spectra for polymeric condensed tannins and furthermore the spectra are difficult to interpret. Solid state NMR spectroscopy has been used in a study [51]. The results obtained by this technique agree well with those obtained with the butanol-HCl method. The NMR spectroscopy is an expensive technique and requires special expertise, and therefore cannot be a routine technique. Although oligomers and flavonoid monomers can be separated by HPLC, condensed tannins cannot be resolved by HPLC. No method based on HPLC is available for reliably quantifying condensed tannins in feedstuffs. For ecological and nutritional studies that may include hundreds of samples it is impractical to attempt to characterize and quantitate each individual compound. Instead, a broad screening methods based on the similar chemistries of groups of related compounds could provide quantitative information about classes of compounds. This strategy has been employed in many studies of condensed tannins (under appropriate conditions, degradation of condensed tannins to anthocyanidins and then its determination spectrophotometrically). The same approach could be employed for hydrolysable tannins (see above) and must be employed. As is evident from the following sections that leaves from most browses and trees contain both condensed and hydrolysable tannins but the determination of hydrolysable tannins has been neglected in the past, and without quantifying hydrolysable tannins or at least screening for its absence the adverse effects of tannins have been attributed to the condensed tannins.

From the above discussion, it is evident that each type of tannin responds differently in each of these assays. This variability in response makes it impossible to use any single method. In order to qualitatively and quantitatively describe the tannins in an extract, we have used a battery of tannin assays. In our work, we used an array of methods. The first method was the Folin-Ciocalteu method in combination with PVPP [2] to measure extractable total phenols and total tannins. Tannic acid from Merck was used as a standard in this procedure. The reasons for using tannic acid from Merck were that the values obtained by the spectrophotometric and the gravimetric methods agree well [2] and there does not seem to be batch-to-batch variation in tannic acid from this source [49]. Although tannic acid obtained commercially is a poorly defined substance, the above advantages and its easy availability and inexpensiveness make it suitable for use as a standard for determination of total phenols and total tannins. For further discussion in this area, the readers are referred to [5, 49]. The second method used in our studies is the butanol-HCl-iron method [11] for determination of condensed tannins, although this assay does not appear to predict the biological effects of tannins (see below). Nevertheless, it gives a good indication of the extent of presence of condensed tannins. Using this method we have expressed the results as leucocyanidin equivalent using the extinction coefficient. For studies on tree leaves, we believe that the use of an absolute standard (purified tannin preparation obtained using Sephadex LH-20) in the butanol-HCl-iron assay is not appropriate due to presence of hydrolysable tannins as impurities in this preparation which will overestimate condensed tannins in a feedstuff [18]. It is pertinent to mention that the difference between the total tannin value obtained by the procedure [2] and the condensed tannin value by the butanol-HCl-iron method is not a measure of hydrolysable tannins, due to different principles of these methods and different standards used (tannic acid is a hydrolysable tannin and can not be used as a standard in the butanol-HCl-iron assay). The third method was the determination of hydrolysable tannins using the rhodanine or HPLC method. The fourth method used in our laboratory is one of the protein precipitation capacity methods for tannins generally [37] or radiolabelled BSA method [35]. For laboratories where laboratory facilities are limited, the radial diffusion method [40] for determination of protein precipitation capacity could also be used. Lately, we have also incorporated the bioassay (based on inactivation of tannins by addition of a tannin-complexing agent, polyethylene glycol in the *in vitro* Hohenheim gas method, and measurement of an increase in gas and change in microbial mass production as a measure of effects of both extractable and bound tannins) into an array of methods. This bioassay helps in predicting effects of tannins on only rumen fermentation. There is a need to develop an *in vitro* approach for measurement of post-rumen effects of tannins.

The suggested procedures for collection, drying, storage and grinding of sample; tannin extraction; and tannin assays are available in this manual.

1.2.8. Screening of condensed and hydrolysable tannins and characterization of phenolic by thin layer chromatography

Two dimensional thin layer chromatography using the cellulose plates, solvent systems comprising of 2 ml glacial acetic acid and 98 ml water followed by another solvent having 60 ml n-butanol, 15 ml glacial acetic acid and 25 ml water, and spray reagents (vanillin-HCl, ferric ion, potassium iodate and sodium nitrite) can be used to screen plants for hydrolysable and condensed tannins. The vanillin-HCl reagent detects flavan-3-ols, e.g., catechin and epi-catechin, and condensed tannins; the ferric ion reagent all phenolic compounds, tannins and nontannins; the potassium iodate reagent gallic acid and its esters, i.e. gallotannins; and the sodium nitrite reagent ellagic acid and its esters. Before spraying the plates with the reagents, the UV-light could also be used to detect some of the phenolic compounds on the plate. Gallotannins and galloyl esters give violet fluorescent spots, the intensity of which increases on fumigation with ammonia vapours. Ellagic also produce violet spots, which get darkened on exposure to ammonia vapours.

Condensed tannins can be characterised using the cellulose plates using a solvents system of 100 ml of 85% formic acid, 12 M HCl and 30 ml water, followed by drying the plate, turning it 90 degrees and then running in the second solvent comprising of 20 ml n-pentan, 10 ml glacial acetic acid and 10 ml water. The condensed tannins in the plant extracts are hydrolysed at approx. 95°C for 60 min using the butanol-HCl reagent. The aliquots from this are spotted on the plate and the plate is run in the above two solvents. The anthocyandins are detected using different spray reagents and characterized from their Rf-values and running the anthocyanidin standards such as chlorides of cyaniding, delphinidin, pelargonidin, etc along with the sample. The preparative thin layer chromatography plates (thickness: 0.5 to 1 mm) could also be used for separation and isolation of tannins.

Detailed thin layer chromatography plate procedures are available in this manual.

1.3. Level and activity of different types of tannins in feedstuffs

1.3.1. Leaves from trees and browses

Information on tannin level and activity in samples of leaves from some trees and shrubs of different regions has been presented in [52–54]. The salient results from the study [52] are presented in Table I. The trees were from sub-humid, tropical regions of Benin and arid and semi-arid regions of Zimbabwe and Niger in Africa, and from sub-tropical regions in the foot-hills of the North-West Humid Himalayan region. There was no significant difference in the crude protein (CP) contents of forages obtained from the Himalayan and African regions, however the levels of extractable total phenols (TP) and the biological activity of tannins expressed as the protein precipitation capacity (PPC) were significantly higher for the African forages. The operational activity of tannins expressed as mg protein precipitated per unit of phenols was also significantly higher in forages from the African regions. For a small set of leaves from arid and semi-arid zones of the Middle East and India TP, PPC, and tannin activity were closer to those for the African forages (Table I). The climate of the African regions from which the forages were collected is harsh and availability of water is also lower than in foot-hills of the Himalayan region, which could possibly explain higher expression of tannins in the African samples. This finding is corroborated by the higher levels and activities of tannins found in forages from arid and semi-arid zones in the Middle East and India. The African plants collected for this study, and especially their bark, could possibly be exploited as a source of vegetable tannins for various industrial applications including the tanning of leather. Given the large differences in the levels of TP, PPC, and tannin activity observed between the two regions, it is tempting to postulate that the antinutrients in forages from trees and shrubs would be of lesser concern in sub-tropical regions of developing countries than in the hot and dry regions of Africa. Various other observations in the literature have also shown higher levels of defensive compounds in plants grown under harsh environmental conditions and thus support the above hypothesis.

Table I Crude protein, total phenols, protein-precipitating capacity and tannin activity in leaves of trees and shrubs from African and Himalayan (foot-hills) regions (data are on DM basis)

	Crude protein	Total phenols (%; as tannic acid equivalent)	Protein precipitation capacity (mg BSA pptd./g)	Tannin activity (mg BSA pptd./ mg phenol)
Himalayan region				
Mean	15.2	6.0^{a}	56.0^{a}	0.66^{a}
SE (n = 16)	1.2	1.0	15.9	0.17
African regions				
Mean	14.1	15.7^{b}	327.2^{b}	1.97^{b}
SEM (n = 13)	1.2	4.3	113.6	0.47
Other arid and semi-arid zones				
Mean	11.3	10.6^{a,b}	245.5^{a,b}	2.30^{b}
SEM (n= 7)	0.97	0.5	24.5	0.17

Mean values for the three groups with different superscripts in a column differ significantly (P <0.05), The data for this table have been extracted from reference [52].

The tannins and other nutritional parameters of several browse plants from the central zone of Delta State, Nigeria [53] showed that the browses (*Cassia siame*, *Vernonia amygdalina* and *Gliricidia sepium*) were low in extractable tannins (tannins, T as tannic acid equivalent < 1.8%, and CT as leucocyanidin equivalent ≤ 0.1%) and high in crude protein (15–28%) suggesting these browses have high potential for the nutrition of small ruminants in the region.

The phenolic compounds, PPC, and *in vitro* gas from leaves of 37 tropical browses collected from the International Livestock Research Institute forage seed multiplication centre, Zwai (Ethiopia) showed considerable variation among the browse species in the contents of CP and phenolic compounds. The CP content ranged from 5.4% in *Eucalyptus microtheca* to 27% in *Sesbania hirtistyla*. The CT, TP, and T contents ranged from 0 to 26.3%, 1.8 to 25.3%, and 0.2 to 21.4% respectively. Among the browse species, *Acacia hockii* was the highest in TP and PPC. The *in vitro* gas production on incubation of 500 mg DM ranged from 11.2 to 112.1 ml and from 23.6 to 118.8 ml in the absence and presence of PEG respectively. The percent change in *in vitro* gas production in the presence of PEG (GV) varied from 0 in *Albizzia schimperiana* to 457 in *A. anguistissima*. Highly significant (P<0.001) correlations were observed between PPC and phenolic contents (r = 0.87, 0.83 and 0.73; n = 37 for TP, T and GV respectively). The percent change in gas production in the presence of PEG was significantly correlated with TP and T (r = 0.76 for both TP and T). The relatively weak correlation between CT and PPC (r = 0.41, n = 37), and between CT and percent increase in gas production could be due to the variation in structure and biological activity of tannins. Good correlation between total phenols and lysozyme inhibition based protein binding assay (r = 0.96) and poorer correlations (r = 0.75) with the butanol-HCl assay have also been reported [55]. The lysozyme inhibition assay and the butanol-HCl assay also had poor correlation (r = 0.39). Furthermore, significant correlations were observed between TP or T and per cent increase in *in vitro* degradability of nitrogen (r = 0.84 and 0.91 respectively) and a poor correlation between CT and percent increase in degradability of nitrogen (r = 0.52) [46]. The CT values by butanol-HCl method do not appear to reflect the biological activity. The PPC of browses and/or the increase in gas production in the rumen simulation system on addition of PEG are considered as a measure of biological activity of tannins [34, 41]. The significant correlations between TP or T and PPC or the increase in gas production on addition of PEG suggest that TP and T could be taken as a measure of biological activity of tannins. It should be noted that these parameters (TP and T) are easier to measure than the PPC and the increase in gas on the addition of PEG. The linear relationship between GV, TP, T, and PPC, is summarized in the following equations:

i) $GV = 9.81TP - 41.9$

ii) $GV = 11.97T - 24.9$

iii) $PPC = 27.1TP - 117.3$

iv) $PPC = 31.5T - 58.3$

At zero GV (equation i and ii), the TP and T will be 4.3 and 2.1% respectively. Similarly at zero PPC (equation iii and iv), the TP and T will be 4.3 and 1.9% respectively. From these relationships, it can be concluded that tropical browses with extractable total phenol and total tannin contents of approximately 4.5% and 2.0% respectively will not produce significant adverse effects on ruminant livestock.

In a subset from the above 37 samples, it was found that the values obtained in the year 2000 were significantly lower compared to the values obtained in 1999, suggesting the analysis for tannins should be conducted as soon as possible, after collection of sample from the field.

The above 37 samples and another from the same region were also analysed for gallotannins using the rhodanine based spectrophotometer [26] and the HPLC method described above [26]; described in this manual). Out of the 38 samples, 21 having the free gallic acid concentration of <10 mg/100 g, and two samples (*A. horrida* and *A. albida*) having a level up to 30 mg/100 g by the HPLC could not be detected by the spectrophotometric method, suggesting that the spectrophotometric method is less sensitive. It is also interesting to note that three samples (*A. saligna, A. tortillas* and *E. macrophylla*) having free gallic acid levels of the same order (28.3, 42 and 47.5 mg/100g) by the HPLC method as in the two samples above showed high values of free gallic acid (150.3, 149.1, 626.9 mg/100 g) by the spectrophotometric method. These higher values of free gallic acid for these three samples and in general higher values for the free as well bound gallic acid by the spectrophotometric method showed that this method is not very specific. It measures other moieties in addition to gallic acid. Twenty-four samples had gallotannin (as gallic acid equivalent) levels of < 0.1%; nine sample between 0.1 and 0.5%; one sample (*Dichrostachys cinerea*) 0.73%; two sample (*A. giraffae* and *Calliandra calothyrsus*) 2 and 1.6% respectively; one (*E. macrophylla*) 3.6%; and one sample (*A. hockii*) 14%.

In winter, in the Himalayan regions, farmers have no option but to feed oak leaves to livestock as no other green fodder is available. However, feeding oak leaves, which are known to be rich in tannins, decreases animal performance and sometimes leads to oak poisoning resulting in the death of animals [56]. The changes that occurred in the various types of tannins in five Himalayan oak species (*Quercus incana, Q. semecarpifolia, Q. ilex, Q. glauca and Q. serrata*) during maturation of the leaves suggested that the contents of hydrolysable tannins, which are known to be toxic, are highest in the young leaves [7, 9]. Hydrolysable tannins contributed substantially more to the protein precipitation capacity of oak leaf tannins than condensed tannins [39]. Other main findings of these studies were: i) the level and activity of tannins differed substantially in the leaves of different oak species being highest in *Q. serrata* and lowest in *Q. ilex*, ii) the levels of condensed tannins increased with maturity, iii) the activity of tannins expressed as protein precipitation capacity followed a pattern similar to that observed for hydrolysable tannins being higher in young leaves and becoming lower with maturity, and iv) the relative degree of polymerisation of condensed tannins increased with the maturity of the leaf suggesting that they turn into tannins of higher molecular weight.

1.3.2. Fallen oak leaves

In the Western and Central Himalayas, the most widely available oak species is *Q. incana*. In the month of May the mature leaves fall from the trees and these form a huge biomass in the forest. The levels of total phenols and condensed tannins and the protein precipitation capacity of tannins in fallen leaves was about 70% lower than in fresh ('on-the-tree') leaves. Senescence processes decrease the contents of both hydrolysable and condensed tannins, and also polymerise the condensed tannins to higher polymers that seem to loose capability to bind proteins [18]. Although tannin levels and their activity were substantially lower in the fallen oak leaves, their *in sacco* dry matter digestibility was low (13.5%). Fallen oak leaves do not seem to have a potential as a feed component.

1.3.3. Weeds

Weeds are generally considered to be undesirable plants as they can cause yield losses but 'fodder weeds' are traded as a valuable feed in Niger in Africa. The weeds, *Alysicarpus ovalifolius, Commelina forskalali, Eragrostis tremula* and *Zornia glochidiata* which are sold in the market were found to contain negligible amounts of tannins (Table II) and some reasonably high levels of crude protein (4.7 to 31.7%; the higher the content of crude protein, the higher the market price). The calculated metabolizable energy values of these weeds were from 5.4 to 8.8 MJ/kg DM. Some weeds might have medicinal value in curing animal illnesses [57]. The role of weeds may thus be more important than presently perceived. Agricultural research programmes should be encouraged to consider the use of fodder weeds to improve both the quality and quantity of feed in the West African Sahel.

1.3.4. Legume straws and foliage

Legume seeds are generally rich in tannins. A study [58] on legume straw and foliage samples of *Lathyrus sativus, Vicia narbonensis, V. sativa* and *V. ervilia* showed that the tannin levels of legume straws are negligible and those of legume foliage with pods sampled in the dough stage are very low (Table II), but that this does not appear to affect adversely their nutritive value. The foliage samples had higher protein, organic matter digestibility and metabolizable energy compared to their straws, and that of legume straws had higher protein, organic matter digestibility and metabolizable energy compared to cereal straws. The legume foliages are lower in crude protein compared to fresh forages of alfalfa, white clover and berseem, but have comparable organic matter digestibility and metabolizable energy [58]. The legume foliages and straws appear to hold promise for incorporation into livestock diets, especially in West Asia and North Africa where the major limiting factor to livestock production is the inadequate feed supply. Shortages can be alleviated by the introduction of leguminous forage species in the existing fallow lands.

1.3.5. Faba beans

The European Union is heavily dependent on the import of protein-rich feedstuffs, especially of soya bean meal from North and South America. To cover the gap between demand and supply of protein-rich feed components, lately use of faba beans (*Vicia faba*) as a source of animal feed has gained considerable interest in Western Europe. Several cultivars of *Vicia* have been produced under various breeding programmes. The knowledge of quality attributes for these cultivars is imperative for breeders to be able to produce cultivar(s) of required characteristics.

Nutrient and antinutrient contents of six colour-flowering and six white-flowering cultivars of *V. faba* bean were measured [59] revealing that the beans had a high protein content but were low in sulfur-containing amino acids. The organic matter digestibility, metabolizable energy and *in vitro* rumen nitrogen degradability of white-flowering cultivars were significantly higher than those of colour-flowering cultivars. The values for the white-flowering cultivars were similar to those for soyabean meal. The main factor, which could limit the use of colour-flowering cultivars of *V. faba* beans in the diets of monogastrics, is the level of tannins (Table II), which may depress protein utilization.

It is evident from Table II that tannin levels are low (tannins $< 1.4\%$ as tannic acid equivalent; condensed tannins $< 2.6\%$ as leucocyanidin equivalent; increase in gas on addition of PEG $< 4\%$) in weeds, foliages and straws of legumes, and colour-flowering cultivars of faba beans. These low levels of tannins may be beneficial for ruminants because tannins protect dietary protein from degradation in the rumen, and increase the efficiency of microbial protein synthesis in the rumen [60, 61] thus leading to a higher supply of amino acids to the post rumen of the host animal.

Table II Tannin levels in various unconventional feedstuffs (weeds, legume straws and foliage and faba beans, values are mean ± SD

	Total phenols[a] %	Tannin[a] %	Condensed tannins[b] %	Tannin bioassay[c] (% increase in gas)
Weeds[1]				
Alysicarpus Ovalifolius	—	0.6–0.8	0.05–0.15	—
Commelina forskalali	—	0.4–2.6	0.05–2.30	—
Eragrostis tremula	—	0.6–3.8	Not detected	—
Zornia glochidiata	—	0.3–0.4	Not detected	—
Legume straws and foliage[2]				
Straws				
Lathyrus sativus (n = 5)	1.2 ± 0.21	0.42 ± 0.07	0.26 ± 0.22	1.1 ±1.1
Vicia narbonensis (n = 9)	0.98 ± 0.15	0.37 ± 0.09	0.13 ± 0.05	0.8 ± 1.5
Vicia sativa (n=10)	1.14 ± 0.4	0.52 ± 0.27	0.51 ± 0.53	1.8 ± 2.1
Vicia ervilia (n=1)	0.65	0.18	0.47	1.1
Foliages				
Vicia sativa (n=14)	2.34 ± 0.7	1.35 ±0.62	1.72 ± 0.85	4.2 ± 1.6
Vicia narbonensis (n = 1)	1.37	0.6	0.43	0
Faba beans[3]				
White flowering (n=6)	0.45 ± 0.04	0.01 ± 0.006	Not detected	—
Colour flowering (n=6)	2.01 ± 0.57	1.41 ± 0.44	2.62 ± 0.71	—

[a] as tannic acid equivalent; [b]as leucocyanidin equivalent using butanol reagent; [c]percent increase in gas on incubation of sample in the presence and absence of polyethylene glycol in *in vitro* gas method.
[1,2,3] from references [57–59] respectively.

1.3.6. Agro-industrial and forestry by-products

Investigations into various agro-industrial and forestry by-products of Asian origin [8] showed that levels of total phenols and condensed tannins are quite low in pods of *Prosopis juliflora* and *Pitthecolobium saman*, cakes of *Guizotia abyssinica*, *Pongamia glabra*, *Schleichera oleosa*, *Hevea brasiliensis* and *Hibiscus cannabinus*, *Cassia tora* seeds and *Panicum miliaceum* bran. The activity of tannins as capacity to precipitate proteins was not detected in these by-products (Table III). Data on tannins as tannic acid equivalent, presented in Table III for other unconventional

Table III Tannin levels in unconventional feedstuffs (agro-industrial and forestry by-products) (data have been extracted from reference [8])

	Total phenols[a] (%)	Condensed tannins[b] (%)	Protein precipitation capacity (mg BSA pptd./g)
Acacia nilotica pods	34.8	3.6 (3.3)	268.7
Prosopis juliflora pods	2.9	0.7 (1.1)	Not detected
Madhuca indica seed–cake	7.1	8.0 (35.0)	155.7
Mangifera indica seed–kernel	7.5	0.7 (1.3)	166.5
Cassia tora seeds	2.7	0.5 (0.5)	Not detected
Guizotia abyssinica cake	2.2	0.4 (0.3)	Not detected
Pongamia glabra cake	0.7	0.4 (0.4)	Not detected
Schleichera oleosa cake	0.7	0.3 (0.1)	Not detected
Pithecolobium saman pods	0.7	0.3 (0.1)	Not detected
Panicum miliaceum polish	15.3	12.4 (62.2)	243.4
Panicum miliaceum Bran	1.4	0.3 (0.2)	Not detected
Hibiscus cannabinus cake	0.7	0.3 (0.7)	Not detected
Garcinia indica cake	4.5	4.1 (12.9)	34.7
Hevea brasiliensis seed–cake	2.3	0.6 (0.4)	Not detected
Theobroma cacao pods	3.9	3.9 (23.7)	88.4

[a]as tannic acid equivalent; [b] as catechin equivalent using vanillin assay (values in parentheses are as bisulphited quebracho equivalent using acid butanol reagent); BSA, bovine serum albumin.

feedstuffs are not available for these by-products as the method for tannins determination using polyvinylpolypyrrolidone [2] was not developed at the time of conducting these studies. These by-products may be considered safe for incorporation in livestock feed as far as tannins are concerned, and provided other deleterious factors are absent.

Acacia nilotica pods, *Madhuca indica* seed-cake, *Mangifera indica* seed kernel, *Panicum miliaceum* polish, *Garcinia indica* cake and *Theobroma cacao* pods were found to have high content of total phenols and condensed tannins. Protein precipitation capacity of tannins from this source was also very high (Table III). It may be noted that the magnitude of difference for condensed tannins observed by the two methods of condensed tannin (the vanillin and the butanol-HCl assays) is not same. For example for *A. nilotica* pods, *T. cacao* pods and *G. indica* cake, condensed tannin contents by the vanillin assay were 3.6, 3.9 and 4.1% respectively and by the acid butanol assay these values were 3.3, 23.7 and 12.9%. This difference is due to different functional groups of condensed tannins involved in these two assays (see above). Furthermore, different levels of condensed tannins observed by these two methods are also contributed by different standards used. Use of different standards in different methods can also lead to values for condensed tannins which are higher compared to total phenols or total tannins, as is for *Madhuca indica* seed-cake. Advantage of using an array of tannin assays is also evident from the data of *Mangifera indica* seed-kernel. Condensed tannin levels of 0.7 or 1.5% obtained by the two methods in this by-product may be considered as low. Total phenol content of 7.5% as tannic acid equivalent and a high protein precipitation capacity of 166.5 mg BSA pptd./g suggest the presence of high levels of hydrolysable tannins with high affinity towards protein and possibly high biological activity. High protein precipitation capacity of by-products such as pods of *A. nilotica*, *Madhuca indica* seed-cake, *Mangifera indica* seed kernel, *P. miliaceum* polish, *G. indica* cake and *T. cacao* pods suggest that these by-products should be used with caution.

Since agro-industrial and forestry by-products and oak leaves are available in large quantities in India and other South East Asian Countries and could be important sources of protein for animals, several methods for detanninification were developed.

1.4. Effects and fate of tannins in ruminants

General effects of tannins, for example, decrease in *in vivo* nutrient utilisation and in particular protein utilisation, decrease in growth, decrease in palatability and feed intake or decrease in various enzyme activities, etc., are not discussed, since these aspects have been dealt with in a great detail in a number of previous reviews.

1.4.1. Digestion kinetics

Purified tannins from various sources affected (decreased) the rate of gas production far more than the ultimate plateau value, i.e., extent of gas production or the *in vitro* true digestibility [60]. Therefore, the threshold level at which the rate is affected will be lower as compared to the digestibility. One of the possible reasons for this could be that tannins decrease the attachment of microbes to feed particles

[62]. McAllister *et al.* [63] have also shown that *Lotus corniculatus* condensed tannins caused a considerable detachment of *Fibrobacter succinogenes* S85 from colonized filter paper after a 30 min exposure). This could also be due to increase in the growth of microbes, which did not bind to tannins and were left active in the system. The adaptation of microbes to tannins with a delay in degradation of tannins does not appear to be responsible, at least for condensed tannins (as these were not degraded by rumen microbes [60, 64]), for the similar potential extent of digestion. The lower intake of tannin-rich feeds is generally attributed to their astringent taste. Besides this unpleasant taste, the lower rate of digestion (higher rumen fill) in presence of tannins [60] could also be responsible for the lower feed intake. The results of [65, 66] also suggest lower *in vivo* rate of feed digestion in presence of condensed tannins.

Purified tannins from leaves of different trees even at the same concentration produced effects of different magnitudes in the rate of gas production and true digestibility. Furthermore, quebracho tannin (a condensed tannin) had a greater effect than tannic acid (a hydrolysable tannin) [60], suggesting the importance of conducting studies on the relationship between tannin structure and activity.

The measurement of gas alone in *in vitro* systems could lead to misleading conclusions on the effects of tannins, or potential rumen modulators on the extent of digestion of tanniniferous feeds and microbial protein synthesis in the rumen. This would adversely affect the evaluation of supplementation strategies and breeding programmes for developing varieties and cultivars of good nutritional value [44, 61]. For tannin-free feeds, gas measurement can be combined with the residue determination (truly degraded substrate), but for tannin-rich feeds, determination of true and apparent digestibility by the use of VanSoest's detergent system of fibre analysis, proved to be meaningless due to the appearance of tannin-protein complexes in the true as well as the apparent residues [67]. For tannin-rich feeds, it is therefore suggested that gas measurements should be combined with the determination of microbial mass using internal or external markers.

1.4.2. Antinutrient-antinutrient interactions

Studies on antinutrient-antinutrient interactions with special reference to tannin-saponin interactions revealed that the effects of both tannins and saponins on decrease in apparent and true digestibilities and gas production were additive, and the hypothesis that "the simultaneous presence of tannins and saponins might alleviate the adverse effects of each other", does not hold good for the tannins (tannic acid and quebracho tannins) and saponins (Quillaja saponins) [60]. The results could be different for saponins and tannins from other sources, which should be further explored. The biological effects of both tannins and saponins are dependent on their chemical properties. The results of such studies may extend the understanding of the nutritional role of complex mixtures of antinutrients in feeds, as generally, antinutrients do not exist in isolation. Furthermore, grazing animals interact with a number of antinutrients daily. Some other related aspects worth mentioning are: i) presence of saponin and tannin in the diet of mice has been shown to abolish adverse effects of both these factors [68], ii) interactions between lectins and tannins removed the inhibitory action of tannins on amylase [69], and iii) tannins may interfere with the release of HCN from the hydrolysis of cyanogens thus reducing their deleterious effects in certain ecological situations [70]. Very little is known in the area of 'antinutrient-antinutrient' interactions. Future studies need to be directed in this field which are likely to have immense nutritional and ecological significance.

1.4.3. Biological significance of bound condensed tannins

A substantial amount of tannins could be present in the bound form [71]. In fresh ('on-the-tree') leaves, tannins do not appear to be present in the bound form [72]. These are predominantly formed while sample preparation. The study of the biological significance of bound condensed tannins has always been a challenge. The nutritional significance of condensed tannins bound to neutral detergent fibre was studied by a combination of different approaches. 1) The correlation between protein binding capacity of condensed tannins-rich neutral detergent fibre and their condensed tannin content was very low ($r = -0.21$). 2) The condensed tannins-rich neutral detergent fibre did not affect growth of a tannin-sensitive strain of *Clostridium perfingens*. 3) Addition of PEG to the incubation medium containing rumen microbes and neutral detergent fibre rich in bound condensed tannins increased the gas production up to 100% [73]. These results suggest that bound condensed tannins seem to be inert and do not affect microbial fermentation as long as they remain bound but once they are released into the medium as a result of microbial action they can affect microbial fermentation. The effect of bound tannins on digestion is not necessarily related in any way to that of free tannins. The microbes, in particular fibre-digesting microbes may encounter high levels of these tannins in their vicinity during digestion. Presently, no method exists to determine what concentrations of tannins actually exist in the vicinity of microbes during digestion and what implications bound tannins have on the digestion kinetics of feeds.

1.4.4. Efficiency of microbial protein synthesis

Microbial protein synthesis *in vitro*, expressed as ^{15}N incorporation into microbes per unit of short-chain fatty acid production was more efficient in the presence of tannins [60]. Although tannins decreased the availability of nutrients, they did cause a shift in the partitioning of nutrients so that a higher proportion of available nutrients were channelled to microbial mass synthesis and less to short-chain fatty acid production [61]. The higher animal performance observed when the diet contains low levels of tannins has generally been attributed to the protection of feed protein from degradation in the rumen, leading to increase in the flux of essential amino acids to small intestine and increase in the absorption of essential amino acid to blood [74]. It may be noted that these beneficial effects of tannins have been demonstrated using temperate forages, and little evidence exists that tropical forages with low levels of condensed tannins, probably because of their higher activity, have beneficial effects mediated through higher absorption of essential amino acids. The studies [60, 61] showed that these beneficial effects of tannins *in vivo* could also be due to higher efficiency of microbial protein synthesis in the rumen. The conclusion that efficiency of microbial protein synthesis is higher in the presence of tannins was arrived at using two different approaches: i) by using purified tannins [60], and ii) by using polyethylene glycol, a tannin-complexing agent [61], the presence of which decreased the efficiency of microbial protein synthesis. The latter [61] opened up avenues for optimising the release of energy producing nutrients from tannin-rich leaves in the presence of polyethylne glycol in order to increase the efficiency of microbial protein synthesis. Polyethylene glycol has the potential to increase the efficiency of utilization of diets containing high amounts of tannins (see above), however, the addition of this tannin inactivating agent, although increases the availability of nutrients from the feed, actually decreases the efficiency of their utilization for microbial protein synthesis probably because of spillage or uncoupling of released energy. The efficiency of microbial protein synthesis under such a situation could be enhanced through proper supplementation strategies [75].

From the above discussion, it is evident that tannins at low levels have the potential to modulate rumen fermentation towards maximizing microbial protein synthesis. The decrease in the rate of digestion of feeds by tannins could help synchronizing the release of various nutrients, which in turn might be responsible for increases in microbial efficiency. Studies are required to know the levels of tannins in order to have this positive response. The concentration of tannins should not be high enough so that the true digestibility of the substrate is appreciably decreased [60]. At these high concentrations of tannins, the advantage provided by the higher efficiency of microbial protein synthesis (higher proportion of truly degraded substrate leading to microbial mass synthesis) will be offset by the absolute lower amount of truly degraded substrate. Another factor which will also influence the microbial nitrogen supply to the lower intestine could be a greater proportion of dry matter which may be digested *in vivo* as a consequence of a slower rate of digestion or dry matter passage through the rumen, despite the reduced microbial population or their activity [76]. Increase in the efficiency of microbial protein synthesis and decrease in the protein degradability of feed protein in the rumen are beneficial for ruminants as they increase the supply of non-ammonia nitrogen to the lower intestine for production purposes, resulting in higher milk, meat and wool production. In addition, these effects lead to protein-sparing effects in ruminants and reductions in methane production and nitrogen excretion to the environment, thereby reducing emission of environmental pollutants besides producing more meat, milk and wool. Feeding strategies need to be designed to exploit these beneficial effects.

The higher molar proportion of propionate in the *in vitro* fermentation system and lower protozoal counts produced by tannins [60, 64] are consistent with the higher efficiency of microbial protein synthesis observed in presence of tannins. Although effects, *in vivo,* of tannins on rumen protozoal counts and molar proportions of short-chain fatty acids are inconsistent [65, 77], some evidence exists for lower protozoal number [78] and higher molar proportion of propionate [65] in presence of tannins. Recently, it has been shown that Calliandra tannins at a level of 2 to 3% in diet reduced fibre degrading bacteria but the efficiency of microbial protein synthesis was not affected. Similarly *Leucaena leucocephala* and leucaena hybrid KX2 tannins at levels of 7.3 and 11.6% respectively in the diets did not affect microbial protein flow as estimated by excretion of urinary purine derivatives [79, 80]. The generally higher efficiency of digestion systems containing tannins could be due to specific inhibition or enhancement of a group(s) of microorganisms. This hypothesis is under investigation using 16S rRNA targeted oligonucleotide probes.

1.4.5. Fate of tannins

There is sufficient evidence in literature suggesting degradation of hydrolysable tannins by rumen microbes (for references see Introduction of [60]). In addition, phenolic glycosides such as rutin and naringin and flavonoids such as quercitrin are metabolised by rumen microbes [81]. For the first time, the study [60] showed that rumen microbes are not capable of degrading condensed tannins. In another study [64], rumen microbes were exposed to small amounts of quebracho tannins using a rumen simulation technique (RUSITEC) for 8 days in order to induce enzymes capable of degrading condensed tannins. But again, there was no degradation of condensed tannins by these 'exposed' microbes, suggesting that rumen microbes do not have the enzymatic apparatus for degrading condensed tannins nor can it be induced even after 8 days of tannin exposure using RUSITEC. Anaerobic microbial degradation of condensed tannins do not seem to take place. In another study, condensed tannins (Sephadex LH-20 purified quebracho and *Acacia nilotica* tannins) were degraded within 7 days under aerobic conditions in artificial fermenters (H.P.S. Makkar; unpublished observations).

The fate of tannins *in vivo* was investigated using a nutrient balance approach. Studies on sheep and goats fed tannin-rich *Acacia saligna* leaves showed virtual absence of extractable tannins in faeces [82]. Similar results have been observed by Terrill *et al.* [83] using the butanol-HCl method, and also by Perez-Maldonado and Norton [84]. These balance studies wherein little free condensed tannins have been observed in faeces should be interpreted with caution with respect to their degradation. The condensed tannins do not get hydrolysed by rumen microbes (41, 60], and additionally [^{14}C] condensed tannin based studies suggested little if any condensed tannin-carbon absorbed to the blood [33, 85]. The free condensed tannins present in the feed get bound to fibre fractions and protein in the gastrointestinal tract and are present in faeces but in the unextractable form. The output of condensed tannins and protein in the acid detergent fibre and acid detergent lignin fractions was substantially higher [82]. Negative digestibilities of these fractions were observed by both sequential and non-sequential approaches [86]. It is also evident that the detergent extraction techniques do not predict or determine the *in vivo* digestibility of cell-wall constituents accurately. This was shown to be due to the presence of tannin-protein complexes as artifacts in fiber fractions, especially in samples of faeces [86]. The extractable tannins bind proteins and are recovered in fibre fractions, explaining negligible presence of extractable tannins in the faecal samples. Similar artifacts make the *in vitro* true and apparent digestibilities invalid for tannin-rich feeds [67]. VanSoest's detergent system for determining cell-wall constituents in forages has several applications in nutritional studies but was found to have little relevance for tannin-rich feeds, and the results obtained using these techniques should be treated with caution. Cell wall digestibility of high-tannin browses should be based on cell wall monosaccharide analysis using GC or HPLC. The presence of nitrogen in the faeces in the form bound to tannins and its use as a manure, and the use of tropical legumes containing tannins as a mulch is expected to release nitrogen slowly in the soil which may be beneficial for long duration pastures and crops. Feeding of a diet containing low levels of tannins to ruminants might also be used to partition the nitrogen in a manner to decrease nitrogen excretion in the urine, thereby reducing a valuable nutrient, nitrogen to the environment and decreasing environmental pollution; and to channel this nitrogen to faeces that can be utilised for crop production when used as a manure.

The presence of tannin-protein complexes in fibre fractions, both *in vitro* and *in vivo*, suggest that not all tannin-protein complexes would get dissociated in the postrumen leading to higher availability of feed protein in the intestine. The reversibility of the tannin-protein complex formation would depend on the binding affinity of tannins to proteins and other macromolecules. The work conducted with *Lotus corniculatus* showed that condensed tannin levels up to 4% in the diet has beneficial effects mediated by protein-tannin complexation at rumen pH and then the dissociation of the complex post-ruminally, enhancing availability of feed protein for production purposes [82, 87]. The reversibility of the complexation process is dependent on the nature of tannins and its affinity towards macromolecules, and hence the generalisation that condensed tannins at approximately 4% in diet, irrespective of the source, would have beneficial effects, must be avoided; since a unit of tannin measured by gravimetric or spectrophotometric methods could have substantially different biological activity [2, 18, 60]. Furthermore, even if the complexation process is completely reversible, the binding of the released tannins to structural proteins and secreted proteins from the intestine could lead to loss of endogenous proteins, and the balance of this endogenous protein loss and the availability of the feed protein at the intestine would determine the beneficial effects of the tannins. Lesions of the gut mucosa have also been observed on feeding tannins [56, 88] which in itself is a wastage of tissue protein and in addition the increased turnover of proteins to repair the mucosal membrane would decrease the availability of amino acids and proteins for productive purposes (milk, meat or wool production and reproduction of animals).

Hanley *et al.* [89] proposed equations for predicting the digestibility of protein and dry matter in ruminants:

$$\text{Digestible protein (g/100 g feed)} = -3.87 + 0.9283X - 11.82Y \quad (I)$$

$$\text{Digestible dry matter (g/100 g feed)} = [(0.9231e^{-0.0451A - 0.03B})(NDF)] + [(-16.03 + 1.02\,NDS) - 2.8\,(P)] \quad (II)$$

where X, crude protein content (6.25 x total N) as percent of dry matter; Y, BSA precipitated (mg/mg of forage dry matter); A, (lignin + cutin) as a percent of the neutral detergent fibre (NDF); B, biogenic silica in percent; NDF, neutral detergent fibre in percent; NDS, neutral detergent solubles in percent (100 minus percent NDF); P, reduction in protein digestion in percent (the 11.82 Y term in the calculation of digestible protein in equation I).

Based on these parameters, the predicted protein and dry matter digestibilities were 34.4 and 45.5% respectively. The measured dry matter digestibility for goats (40.8%) was closer to the predicted value than that of sheep (31.9%) but still the difference is substantial. There was a big difference between predicted and measured protein digestibilities for both species [82]. The difference could be attributed to the presence of proline-rich proteins in the saliva for deer; animals used to develop the predictive equations, but were absent in goats and sheep (see below). The predictive equations proposed in [89] cannot be applied to sheep and goats.

1.4.6. Adaptation to tannins

Very high levels of tannin intake by animals can produce toxicity and can even cause death [56]. The observed adverse effects of a tannin-rich feed do not reflect its toxic potential, but rather the toxicity resulting from remaining tannins or their degraded products not detoxified by the animal's detoxification apparatus. Animals normally consuming tannin-rich feeds appear to develop defensive mechanisms against tannins. In some animals, the salivary proline-rich proteins (PRPs) are considered to be the first line of defence against dietary tannins and therefore less protein is required to bind all the tannins, resulting in a quantitative saving of nitrogen. In addition, PRPs being rich in non-essential amino acids are excreted instead of essential amino acids from dietary proteins, providing a qualitative nitrogen saving. The PRPs have not been found in saliva of sheep, cattle and goats; and cattle fed oak leaves rich in tannins, such as those in the hilly areas of India, Nepal and China during winter, do not appear to produce salivary PRPs (as do rats and deer) as a defence strategy against tannins [90–92]. However, other proteins are present in the saliva of these cattle fed tannin-rich oak leaves, which are not rich in proline but which do have a high affinity for tannic acid and tend to form soluble tannin-protein complexes. The primary role of these cattle salivary proteins does not appear to be to neutralize dietary tannins, as these proteins are also present in cattle reared on a tannin-free diet [92]. Recently, it has been shown that histatins a group of peptides (31 to 38 amino acids, virtually absent in proline and high in histidine), with antibacterial and antifungal activities, are present in human and monkey saliva that precipitate tannins more effectively than PRPs and gelatine, and most of the condensed tannin-protein complexes remained insoluble under conditions similar to those in the stomach and small intestine, suggesting that histatins may act as a defence against dietary tannin in humans [93]. The significance of histatins in adaptation to tannins by livestock is not known.

Microbe(s) capable of degrading condensed tannin-protein complexes were not detected in faeces of tannin-fed cattle [82, 94]. In addition, our *in vitro* studies [60, 64] showed absence of microbes in the rumen capable of degrading condensed tannins. Although microoganisms resistant to high levels of condensed tannins have been reported to be present in ruminants consuming high levels of tannin-containing feeds [95–98 and references in 98], there is no evidence of rumen microorganisms capable of degrading condensed tannins. It therefore appears that cattle fed oak leaves or other tannin-rich feeds do not defend themselves by producing salivary PRPs or by taking advantage of rumen microbes capable of degrading condensed tannins. The microorganisms present might however have some adaptive mechanism, which enables them to degrade hydrolysable tannins faster [99] or decrease activity of hydrolysable tannins through methylation of phenolic hydroxy groups, and these microbes might be capable of working efficiently in high concentrations of tannins. The adaptive mechanisms through which the rumen microbes could tolerate and work efficiently in presence of high tannin levels could be the protection of key membrane proteins by strategic deployment of lipids [55], the secretion of extracellular polysaccharides [98] which has high affinity for binding to tannins and/or through a formation of thick high-tannin affinity glycoprotein [100, 101], and thus prevent tannins to cause adverse effects on rumen microbes. Another possible mechanism could be more efficient liver detoxification apparatus induced by low molecular weight phenolic substances. Since tannins are unlikely to be absorbed and transported to the cells, the activation of liver detoxification enzymes by tannins does not appear to play any significant role in adaptation of animals to tannins. Nevertheless, activation of intestinal detoxification apparatus by tannins is possible. Research to investigate contribution of these mechanisms towards tannin tolerance is warranted. In mid-nineties some work showed potential for neutralising the effects of tannins by transferring microbes from 'adapted' to 'unadapted' animals [97], but probably because of the beneficial effects not being consistently observed, not much progress has been made in practical exploitation of this approach.

1.4.7. Post-absorptive effects

Poisoning from the consumption of oak leaves [56] and yellow-wood toxicity from the leaves of *Terminalia*, *Clidemia,* and *Ventilago* has been attributed to the presence of hydrolysable tannins, in particular gallotannins [See 81]. Rumen microbes are capable of degrading hydrolysable tannins. The toxicity, therefore, appears to be due to absorption of degraded products of hydrolysable tannins and higher load of phenols in the blood stream, which is beyond the capability of liver to detoxify them.

Studies have recently been reported showing absorption of catechins, quercetin, isoflavones diadzein and genistein. The bioavailability of the flavonoids, possibly as the glycoside, ranges from 10 to 50% [102]. The condensed tannins are not absorbed into the blood stream [83], these, therefore, under normal physiological conditions, are not likely to damage organs such as liver, kidney, spleen etc., as has been observed for hydrolysable tannins [56, 103]. PEG when infused into abomasums supported a higher wool production than the same amount of PEG added to the rumen of animals fed Calliandra, suggesting the importance of tannins post-ruminally [104]. Under situation of intestinal damage due to consumption of high levels of tannins or of other intestinal membrane irritants, condensed tannins may get absorbed into blood and can cause organ damages similar to those observed for hydrolysable tannins. Saponins are also known to increase the intestinal permeability [105]. Consumption of feeds containing saponins and tannins might lead to higher absorption of both condensed and hydrolysable tannins, and possibly of nutrients such as glucose and amino acids. It may also be noted that tannins are known to bind saponins [68] and proteins [106] and may even decrease the absorption of tannins and amino acids into the blood. The factors leading to higher or lower absorption of tannins or nutrients in presence of saponins or as a matter of fact of any other phytochemical/allelochemical are not known.

From the above, it is evident that tannins have both beneficial and adverse effects depending upon their nature and the amount an animal consumes. Tannins isolated from leaves of various multipurpose trees and browses also have anticarcinogenic activity [50]. Tannins also protect ruminants from bloat and have anthelmintic effects [107]. Legume tannins could also enhance quality of the silage by preventing excessive degradation of feed proteins. Additionally, the proanthocyanidins, both in free form and bound to proteins, have been shown to have free radical scavenging abilities [108] and to decrease the susceptibility of healthy cells to toxic agents. In light of these observations, the supposed role of tannins as antinutritional factors may need to be revised and it is no longer appropriate to refer to them as antinutritional. The application of the research results on a wider scale is likely to be environmentally beneficial. The spread of multipurpose trees and legume forages leads to improvements in soil fertility and control of erosion and their use as sources of feed for ruminants may result in lower methane production as well as increased milk, meat and draught power. Novel ways of using agro-industrial and forestry by-products as livestock feeds convert "disposal problems" into "opportunities for development". The possibility of natural plant products that can be used as growth promoters for livestock provides cheaper and safer alternatives to synthetic compounds. Some of the research activities and novel methodologies developed have opened up several avenues for future studies, which could have important ramifications in many fields beyond the agricultural sciences.

1.5. References

[1] HAGERMAN, A.E., ZHAO, Y., JOHNSON, S., Methods for determination of condensed and hydrolysable tannins. In: Shahidi, F. (Ed.) Antinutrients and Phytochemicals in Food, ACS Symposium Series No. 662, American Chemical Society (1997) 209–222.

[2] MAKKAR, H.P.S., BLUMMEL, M., BOROWY, N.K., BECKER, K., Gravimetric determination of tannins and their correlations with chemical and protein precipitation methods, J. Sci. Food Agric. **61** (1993) 161–165.

[3] BROADHURST, R.B., JONES, W.T., Analysis of condensed tannins using acidified vanillin, J. Sci. Food Agric. **29** (1978) 788–794.

[4] PRICE, M.L., VAN SCOYOC, S., BUTLER, L.G., A critical evaluation of the vanillin reaction as an assay for tannin in sorghum grain, J. Agric. Food Chem. **26** (1978) 1214–1218.

[5] MAKKAR, H.P.S., BECKER, K., Some problems in the determination of tannins and possible solutions, Acta Horticulturae **381** (1994) 782–788.

[6] MAKKAR, H.P.S., BECKER, K., Vanillin-HCl method for condensed tannins: effect of organic solvents used for extraction of tannins, J. Chem. Ecol. **19** (1993) 613–621.

[7] MAKKAR, H.P.S., DAWRA, R.K., SINGH, B., Changes in tannin content, polymerization and protein-precipitation capacity in oak (*Quercus incana*) leaves with maturity, J. Sci. Food Agric. **44** (1988) 301–307.

[8] MAKKAR, H.P.S., SINGH, B., NEGI, S., Tannin levels and their degree of polymerisation and specific activity in some agro-industrial by-products, Biol. Wastes **31** (1990) 137–144.

[9] MAKKAR, H.P.S., SINGH, B., DAWRA, R.K., Tannin levels in the leaves of some oak species at different stages of maturity, J. Sci. Food Agric. **54** (1991) 513–519.

[10] BATE-SMITH, E.C., Tannins of herbaceous leguminosae, Phytochemistry **12** (1973) 1809–1812.

[11] PORTER, L.J., HRSTICH, L.N., CHAN, B.G., The conversion of procyanidin and prodelphinidins to cyanidin and delphinidin, Phytochemistry **25** (1986) 223–230.

[12] MAKKAR, H.P.S., SINGH, B., Determination of condensed tannins in complexes with fibre and proteins, J. Sci. Food Agric. **69** (1995) 129–132.

[13] MAKKAR, H.P.S., GAMBLE, G., BECKER, K., Limitations of the butanol-hydrochloric acid-iron assay for bound condensed tannins, Food Chem. **66** (1999) 129–133.

[14] TERRILL, T.H., ROWAN, A.M., DOUGHLAS, G.B., BARRY, T.N., Determination of extractable and bound condensed tannin concentrations in forage plants, protein concentrate meals and cereal grains, J. Sci. Food Agric. **58** (1992) 321–329.

[15] PENG, S., SCALBERT, A., MONTIES, B., Insoluble ellagitannins in *Castanea sativa* and *Quercus petraea* woods, Phytochemistry **30** (1991) 775–778.

[16] MATTHEWS, S., MILA, I., SCALBERT, A., POLLET, B., LAPIERRE, C., HERVE DU PENHOAT, C.L.M., ROLANDO, C., DONNELLY, D.M.X., Method of estimation of proanthocyanidins based on their acid depolymerization in the presence of nucleophiles, J. Agric. Food Chem. **45** (1997) 1195–1201.

[17] BUTLER, L.G., PRICE, M.L., BROTHERTON, J.E., Vanillin assay for proanthocyanidins (condensed tannins): modification of the solvent for estimation of the degree of polymerisation, J. Agric. Food Chem. **30** (1982) 1087–1089.

[18] MAKKAR, H.P.S., BECKER, K., Isolation of tannins from trees and shrubs and their properties, J. Agric. Food Chem. **42** (1994) 731–734.

[19] MAKKAR, H.P.S., SINGH, B., KAMRA, K., Biodegradation of tannin in oak (*Ouercus incana*) leaves by *Sporotricum pulverulentum,* Letters in Applied Microbiology **18** (1994) 42–44.

[20] MAKKAR, H.P.S., SINGH, B., Effect of wood ash on tannin content of oak (*Quercus incana*) leaves, Bioresource Technology **41** (1992) 85–86.

[21] MAKKAR, H.P.S., SINGH, B, Effect of steaming and autoclaving oak (*Quercus incana*) leaves on levels of tannins, fibre and lignin and *in-sacco* dry matter digestibility, J. Sci. Food Agric. **59** (1992) 469–472.

[22] MAKKAR, H.P.S., SINGH, B., Detannification of oak leaves: Treatments and their optimization. Anim. Feed Sci. Technol. **36** (1992) 113–127.

[23] MAKKAR, H.P.S., SINGH, B., Effect of storage and urea addition on detannification and *in sacco* dry matter digestibility of mature oak (*Quercus incana*) leaves, Anim. Feed Sci. Technol. **41** (1993) 247–259.

[24] LABARBE, B., CHEYNIER V., BROSSAUD, F., SOUQUET, J-M., MOUTOUNET, M., Quantitative fractionation of grape proanthocyanidins according to their degree of polymerisation, J. Agric. Food Chem. **47** (1999) 2719–2723.

[25] GUYOT, S., MARNET, N., DRILLEAU, J-M., Thiolysis-HPLC characterization of apple proanthocyanidins covering a large range of polymerisation states, J. Agric. Food Chem. **49** (2001) 14–20.

[26] INOUE, K.H., HAGERMAN, A.E., Determination of gallotannins with rhodanine, Anal. Biochem. **169** (1988) 363–369.

[27] MAKKAR, H.P.S., BECKER, K. Gallotannin determination using HPLC, Unpublished.

[28] BATE SMITH, E.C., Astringent tannins of *Acer* species, Phytochemistry **16** (1977) 1421–1427.

[29] HARTZFELD, P.W., FORKNER, R., HUNTER, M.D., HAGERMAN, A.E., Determination of hydrolysable tannins (gallotannins and ellagitannins) after reaction with potassium iodate, J. Agric. Food Chem. **50** (2002) 1785–1790.

[30] LEI Z., JERVIS, J., HELM, R.F., Use of methanolysis for the determination of total ellagic and gallic acid contents of wood and food products, J. Agric. Food Chem. **49** (2001) 1165–1168.

[31] SCHOFIELD, P., MBUGUA, D.M., PELL, A.N., Analysis of condensed tannins: a review, Anim. Feed Sci. Technol. **91** (2001) 21–40.

[32] REED, J.D., HORVATH, P.J., ALLEN, M.S., VAN SOEST, P.J., Gravimetric determination of soluble phenolics including tannins from leaves by precipitation with trivalent ytterbium, J. Sci. Food Agric. **36** (1985) 255–261.

[33] LOWRY, J.B., SUMPTER, E.A., Problems with ytterbium precipitation as a method for determination of plant phenolics, J. Sci. Food Agric. **52** (1990) 287–288.

[34] MAKKAR, H.P.S., Protein precipitation methods for quantitation of tannins: a review, J. Agric. Food Chem. **37** (1989) 1197–1202.

[35] HAGERMAN, A.E., BUTLER, L.G., 1980. Determination of protein in tannin-protein precipitates, J. Agric. Food Chem. **28** (1980) 944–947.

[36] MAKKAR, H.P.S., DAWRA, R.K., SINGH, B., Protein precipitation assay for quantitation of tannins: Determination of protein in tannin-protein complex, Anal. Biochem. **166** (1987) 435–439.

[37] MAKKAR, H.P.S., DAWRA, R.K., SINGH, B., Determination of both tannin and protein in a tannin-protein complex, J. Agric. Food Chem. **36** (1988) 523–525.

[38] HAGERMAN, A.E., BUTLER, L.G., Protein precipitable method for the quantitative determination of tannins, J. Agric. Food Chem. **26** (1978) 809–812.

[39] DAWRA, R.K., MAKKAR, H.P.S., SINGH, B., Protein binding capacity of microquantities of tannins, Anal. Biochem. **170** (1988) 50–53.

[40] HAGERMAN, A.E., Radial diffusion method for determining tannin in plant extracts, J. Chem. Ecol. **13** (1987) 437–449.

[41] MAKKAR, H.P.S., BLÜMMEL, M., BECKER, K., Formation of complexes between polyvinyl pyrrolidone and polethylene glycol with tannins and their implications in gas production and true digestibility in *in vitro* techniques, Brit. J. Nutr. **73** (1995) 897–913.

[42] MAKKAR, H.P.S., BECKER, K., A bioassay for tannins. The XVIIIth International Conference on Polyphenols, Bordeaux, July 15–18, 1996. Polyphenols Communications **96** (1996) 197–198.

[43] MENKE, K.H., STEINGASS, H., Estimation of the energetic feed value from chemical analysis and *in vitro* gas production using rumen fluid, Anim. Res. Develop. **28** (1988) 7–55.

[44] MAKKAR, H.P.S., BLÜMMEL, M., BECKER, K., Potential and limitations of *in vitro* gas method for studying the effects of plant defensive components on rumen fermentation. 3rd International Workshop on Antinutritional Factors in Legume Seeds and Rapeseed, July 8–10 (1998), Wageningen, The Netherlands.

[45] SILANIKOVE, N., PEREVOLOTSKY, A., PROVENZA, F.D., Use of tannin-binding chemicals to assay for tannins and their negative postingestive effects in ruminants, Anim. Feed Sci. Technol. **91** (2001) 69–81.

[46] GETACHEW, G., MAKKAR, H.P.S., BECKER, K.,. Effect of polyethylene glycol on *in vitro* degradability of nitrogen and microbial protein synthesis from tannin-rich browse and herbaceous legumes, Brit. J. Nutr. **84** (2000) 73–83.

[47] JONES, R.J., MEYER, J.H.F., BECHAZ, M., STOLTZ, M.A., An approach to screening potential pasture species for condensed tannin activity, Anim. Feed Sci. Technol. **85** (2000) 269–277.

[48] GOODCHILD, A.V., EL HARAMEIN, F.J., MAKKAR, H.P.S., EL MONEIN, A. A., WILLIAMS, P.C., 18th International Conference on Near-Infrared Spectroscopy, September 15–19 (1997), Haus der Technik, Essen, Germany.

[49] MAKKAR, H.P.S., BECKER, K., Behaviour of tannic acid from various commercial sources towards some chemical and protein precipitation assays, J. Sci. Food Agric. **62** (1993) 295–299.

[50] PERCHELLET, E.M., MOUTASEB, H.U., MAKKAR, H.P.S., PERCHELLET, P., Ability of tannins extracted from various tree leaves to inhibit the biomarkers of tumor promotion in mouse skin *in vivo*, Int. J. Oncology **9** (1996) 801–809.

[51] GAMBLE, G.R., AKIN, D.E., MAKKAR, H.P.S., BECKER, K., Biological degradation of tannins in *Sericea lespedeza* by the white rot fungi *Ceriporiopsis subvermispora* and *Cyathus stercoreus* analyzed by solid state ^{13}C NMR spectroscopy, Appl. Environ. Microbiol. **62** (1996) 3600–3604.

[52] MAKKAR, H.P.S., BECKER, K., Do tannins in leaves of trees and shrubs from African and Himalayan region differ in level and activity? Agroforestry Systems **40** (1998) 59–68.

[53] AREGHEORE, E.M., MAKKAR, H.P.S., BECKER, K., Feed value of some browse plants from Central Zone of Delta State, Nigeria, Trop. Sci. **38** (1998) 97–104.

[54] GETACHEW, G., MAKKAR, H.P.S., BECKER, K., Tropical browses: contents of phenolic compounds, *in vitro* gas production, and

stoichiometrical relationship between short chain fatty acid and *in vitro* gas production, J. Agric. Sci. (*Camb.*) **139** (2002) 341–352.

[55] PELL, A.N., WOOLSTON, T.K., NELSON, K.E., SCHOFIELD, P., Tannins: biological activity and bacterial tolerance. In: Tannins in Livestock and Human Nutrition: Proceedings of an International Workshop, Adelaide, Australia, May 31–June 2, 1999 (Ed., J.D. Brooker), ACIAR Proceedings No. **92** (2000) 111–116.

[56] GARG, S.K., MAKKAR, H.P.S., NAGAL, K.B., SHARMA, S.K., WADHWA, D.R., SINGH, B., Toxicological investigations into oak (*Quercus incana*) leaf poisoning in cattle, Vet Human Toxicol. **34** (1992) 161–164.

[57] LAMERS, J., BUERKERT, A., MAKKAR, H.P.S., OPPEN, M.V., BECKER, K., Biomass production, feed economic value of fodder weeds as by-products of millet, Exp. Agric. **32** (1996) 317–326.

[58] MAKKAR, H.P.S., GOODCHILD, A.V., ABD-EL-MONEIM, A.M., BECKER, K., Cell-constituents, tannin levels by chemical and biological assays and nutritional value of some legume foliage and straws, J Sci. Food Agric. **71** (1996) 129–136.

[59] MAKKAR, H.P.S., BECKER, K., ABEL, H.J., PAWELZIK, E., Nutrient contents, rumen degradability and antinutritional factors in some colour- and white-flowering cultivars of *Vicia faba* beans, J. Sci. Food Agric. **75** (1997) 511–520.

[60] MAKKAR, H.P.S., BLÜMMEL, M., BECKER, K., *In vitro* effects and interactions of tannins and saponins and fate of tannins in rumen, J Sci. Food Agric. **69** (1995) 481–493.

[61] MAKKAR, H.P.S., BLÜMMEL, M., BECKER, K., Application of an *in vitro* gas method to understand the effects of natural plant products on availability and partitioning of nutrients. BSAP Occasional Meeting on "*In vitro* techniques for measuring nutrient supply to ruminants, 8–10 July, (1997), Reading, U.K.

[62] MAKKAR, H.P.S., SINGH, B., NEGI, S.S., Relationship of rumen degradability with biomass accumulation, cell wall constituents and tannin levels in some tree leaves, Anim. Prod. **49** (1989) 299–303.

[63] MCALLISTER, T.A., BAE, H.D., JONES, G.A., CHENG, K.-J., Microbial attachment and feed digestion in the rumen, J. Anim. Sci. **72** (1994) 3004–3018.

[64] MAKKAR, H.P.S., BECKER, K., ABEL, H.J., SZEGLETTI, C., Degradation of condensed tannins by rumen microbes exposed to quebracho tannins (QT) in rumen simulation technique (RUSITEC) and effects of QT on fermentation processes in the RUSITEC, J. Sci. Food Agric. **69** (1995) 495–500.

[65] WAGHORN, G.C., SHELTON, I.D., MCNABB, W.C., Effects of condensed tannins in *Lotus pedunculatus* on its nutritive value for sheep. 1. Non-nitrogenous aspects, J. Agric. Sci. (*Camb.*) **123** (1994) 99–107.

[66] STIENEZEN, M., WAGHORN, G.C., DOUGHLAS, G.B., Digestibility and effects of condensed tannins on digestion of sulla (*Hedysarum coronarium*) when fed to sheep, N.Z. J. Agric. Res. **39** (1996) 215–221.

[67] MAKKAR, H.P.S., BLÜMMEL, M., BECKER, K., *In vitro* rumen apparent and true digestibilities of tannin-rich forages, Anim. Feed Sci. Technol. **67** (1997) 245–251.

[68] FREELAND W.J., CALCOTT, P.H., ANDERSON, L.R., Tannins and saponins:interaction in herbivore diets, Biochemical System Ecol. **13** (1985) 189–193.

[69] FISH, B.C., THOMPSON, L.U. J., Lectin-tannin interactions and their influence on pancreatic amylase activity and starch digestibility, J. Agric. Food Chem. **39** (1991) 727–731.

[70] GOLDSTEIN, W.S., SPENCER, K.C., Inhibition of cyanogensis by tannins, J. Chem. Ecol. **11** (1985) 847–857.

[71] MAKKAR, H.P.S., SINGH. B., Composition, tannin levels and *in sacco* dry matter digestibility of fresh and fallen oak (*Quercus incana*) leaves, Bioresource Technology **37** (1991) 185–187.

[72] GETACHEW, G., MAKKAR, H.P.S., BECKER, K., Tannin-binding agents to alleviate anti-microbial effects of tannins in the rumen. 3rd Tannin Conference 'Plant Polyphenol: Chemistry and Biology', July 20–25 (1998), Bend, Oregon, USA.

[73] MAKKAR, H.P.S., BECKER, K., Nutritional implications of bound proanthocyanidins. XIII Int. Grassland Cong. '97, Winnipeg, Canada, June 9–12 (1997).

[74] WAGHORN, G.C., SHELTON, I.D., Effect of condensed tannins in *Lotus corniculatus* on the nutritive value of pasture for sheep, J. Agric. Sci. *(Camb.)* **128** (1997) 365–372.

[75] GETACHEW, G., MAKKAR H.P.S, BECKER K., Tannins in Tropical browses: Effects on *in vitro* microbial fermentation and microbial protein synthesis in media containing different amounts of nitrogen. J. Agric. Food Chem. **48** (2000) 3581–3588.

[76] MAKKAR, H.P.S., SINGH, B., DAWRA, R.K., Effect of tannin-rich leaves of oak on various microbial activities of the rumen, Brit. J. Nutr. **60** (1988) 287–296.

[77] WANG, Y., DOUGLAS, G.B., WAGHORN, G.C., BARRY, T.N., FOOTE, A.G., PURCHAS, R.W., Effect of condensed tannins upon the performance of lambs grazing *Lotus corniculatus* and lucerne *(Medicago sativa)*, J. Agric. Sci. *(Camb.)* **126** (1996) 87–98.

[78] WANG, Y., WAGHORN, G.C., BARRY, T.N., SHELTON, I.D., The effect of condensed tannins in *Lotus corniculatus* on plasma metabolism of methionine, cystine and inorganic sulphate by sheep, Brit. J Nutr. **72** (1994) 923–935.

[79] MCNEILL, D.M., OSBORNE, N., KOMOLONG, M., NANKERVIS, D., Condensed tannins in the Leucaena genus and their nutritional significance for ruminants. In: Shelton, H.M., Gutteridge, R.C., Mullin, B.F., Bray, R.A. (Eds.), Leucaena – Adaptation, Quality and Farming Systems, ACIAR Proceedings No. **86** (1998) 205–214.

[80] MCNEILL, D.M., KOMOLONG, M., GOBIUN, N., BARBER, D., Influence of dietary condensed tannins on microbial crude protein supply in sheep. In: Brooker, J.D. (Ed.), Tannins in Livestock and Human Nutrition. ACIAR Proceedings No. **92** (2000) 57–61.

[81] MCSWEENEY, C.S., PALMER, B., MCNEIL, D.M., KRAUSE, D.O., Microbial interactions with tannins: nutritional consequences for ruminants, Anim. Feed Sci. Technol. **91** (2001) 83–93.

[82] DEGEN, A., MAKKAR, H.P.S., BECKER, K., BOROWY, N., *Acacia saligna* as a fodder tree for desert livestock. J. Sci. Food Agric. **68** (1995) 65–71.

[83] TERRILL, T.H., WAGHORN, G.C., WOOLLEY, D.J., MCNABB, W.C., BARRY, T.N., Assay and digestion of ^{14}C-labelled condensed tannins in the gastrointestinal tract of sheep, Brit. J. Nutr. **72** (1994) 467–477.

[84] PEREZ-MALDONADO, R.A., NORTON, B.W., The effects of condensed tannins from *Desmodium intortum* and *Calliandra calothyrsus* on protein and carbohydrate digestion in sheep and goats, Br. J. Nutr. **76** (1996) 515–533.

[85] ABIA, R., FRY, S.C., Degradation and metabolism of ^{14}C-labelled proanthocyanidins from carob (*Ceratonia siliqua*) pods in the gastrointestinal tract of the rat, J. Sci. Food Agric. **81** (2001) 1156–1165.

[86] MAKKAR, H.P.S., BOROWY, N., BECKER, K., DEGEN, R., Some problems in fiber determination in tannin-rich forages, Anim. Feed Sci. Technol. **55** (1995) 67–76.

[87] BARRY, T.N., MANLEY, T.R., DUNCAN, S.J., The role of condensed tannins in the nutritional value of *Lotus pedunculatus* for sheep. 4. Site of carbohydrate and protein digestion as influenced by dietary reactive tannin concentrations, Brit. J. Nutr. **55** (1986) 123–137.

[88] DAWSON, J.M., BUTTERY, P.J., JENKINS, D., WOOD, C.D., GILL, M., Effects of dietary quebracho tannins on nutrient utilization and tissue metabolism in sheep and rats, J. Sci. Food Agric. **79** (1999) 1423–1430.

[89] HANLEY, T.A., ROBBINS, C.T., HAGERMAN, A.E., MCARTHUR, C., Predicting digestible protein and digestible dry matter in tannin-containing forages consumed by ruminants, Ecology **73** (1992) 537–541.

[90] JANSMAN, A.J.M., FROHLICH, A.A., MARQUARDT, R.R., Production of proline-rich proteins by the parotid glands of rats is enhanced by feeding diets containing tannins from faba beans (*Vicia faba* L.), J. Nutr. **124** (1994) 249–258.

[91] PEREZ-MALDONADO, R.A., NORTON, B.W., KERVEN, G.L., Factors affecting *in vitro* formation of tannin-protein complexes, J. Sci. Food Agric. **69** (1995) 291–298.

[92] MAKKAR, H.P.S., BECKER, K., Adaptation of cattle to tannins: role of proline-rich proteins in oak-fed cattle, Anim. Sci. **67** (1997) 277–281.

[93] NAURATO, N., WONG, P., LU, Y., WROBLEWSKI, K., BENNICK, A., Interaction of tannin with human salivary histatins, J. Agric. Food Chem. **47** (1999) 2229–2234.

[94] BHAT, T.K., MAKKAR, H.P.S., SINGH, B., Degradation of tannin-protein complexes by microorganisms present in faeces of oak-fed cattle, Letters Appl. Microb. **22** (1996) 257–258.

[95] NELSON, K.E., PELL, A.N., SCHOFIELD, P., ZINDER, S., Isolation and characterization of an anaerobic ruminal bacterium capable of degrading hydrolysable tannins, Appl. Environ. Microbiol. **61** (1995) 3293–3298.

[96] NELSON, K.E., THONNEY, M.L., WOOLSTON, T.K., ZINDER, S.H., PELL, A.N., Phenotypic and phylogenetic characterisation of ruminal tannin-tolerant bacteria, Appl. Environ. Microbiol. **64** (1998) 3824–3830.

[97] MILLER, S.M., KLIEVE, A.V., PLUMB, J.J., AISTHORPE, R., BLACKALL, L.L., An *in vitro* cultured rumen inoculum improves nitrogen digestion in mulga-fed sheep, Aust. J. Agric. Res. **48** (1997) 403–409.

[98] BROOKER, J.D., O'DONOVAN, L., SKENE, I., SELLICK, G., Mechanisms of tannin resistance and detoxification in the rumen. In: Brooker, J.D. (Ed.), Tannins in Livestock and Human Nutrition. ACIAR Proceedings No. **92** (2000) 117–122.

[99] ODENYO, A.A., MCSWEENEY, C.S., PALMER, B., NEGASSA, D., OSUJI, P.O., *In vitro* screening of rumen fluid samples from indigenous African ruminants provides evidence for rumen fluid with superior capacities to digest tannin-rich fodders, Aust. J. Agric. Res. **50** (1999) 1147–1157.

[100] NICHOLSON, K., BUTLER, L.G., ASQUITH, T.N., Glycoproteins from *Colletptrichum graminicola* that bind phenols: implications for survival and virulence of phytopathogenic fungi, Phytochemistry **76** (1986) 3824–3830.

[101] CHIQUETTE, J. CHENG, K.J., COSTERTON, J.W., MILLIGAN, L.P., Effect of tannin on the digestibility of two isosynthetic strains of bird's-foot trefoil (*Lotus corniculatus*) using *in vitro* and *in sacco* techniques, Can. J. Anim. Sci. **68** (1998) 751–760.

[102] KING, R.A., The role of polyphenols in human health. In: Tannins in Livestock and Human Nutrition: proceedings of an International Workshop,

Adelaide, Australia, May 31–June 2, 1999 (Ed., J.D. Brooker), ACIAR Proceedings No. **92** (2000) 75–81.

[103] MCSWEENEY, C.S., KENNEDY, P.M., JOHN, A., Effect of ingestion of hydrolysable tannins in *Terminalia oblongata* on digestion in sheep fed *Stylosanthes hamata*, Aust. J. Agric. Res. **39** (1988) 235–244.

[104] PALMER, B., MCSWEENEY, C.S., Tannins in *Calliandra calothyrsus*: effect of polyethylene glycol (PEG) and an evaluation of 19 accessions. In: Tannins in Livestock and Human Nutrition: proceedings of an International Workshop, Adelaide, Australia, May 31–June 2, 1999 (Ed., J.D. Brooker), ACIAR Proceedings No. **92** (2000) 36–39.

[105] PRICE, K.R., JOHNSON, I.T., FENWICK, G.R., The chemistry and biological significance of saponins in foods and feeding stuffs, CRC Crit. Rev. Food Sci. Nutr. **26** (1987) 27–135.

[106] IKEDO, S., SHIMOYAMADA, M., WATANABE, K., Interaction between bovine serum albumin and saponin as studied by heat stability and protease digestion, J. Agric. Food Chem. **44** (1996) 792–795.

[107] KAHN, L.P., DIAZ-HERNANDEZ, A., Tannins with anthelmintic properties. In: Brooker, J.D. (Ed.), Tannins in Livestock and Human Nutrition. ACIAR Proceedings No. **92** (2000) 130–139.

[108] HAGERMAN, A.E., RIEDL, K.M., RICE, R.E., Tannins as biological antioxidants. In: Plant Polyphenols 2: Chemistry, Biology, Pharmacology, Ecology (eds. Gross, G.G., Hemingway, Yoshida, T.), Kluwer Academic/Plenum Publishers, New York (1999) 495–505.

2. TREATMENT OF PLANT MATERIAL, EXTRACTION OF TANNINS, AND AN OVERVIEW OF TANNIN ASSAYS PRESENTED IN THE MANUAL

2.1. Introduction

Tannins are naturally occurring polyphenolic substances with various molecular weights and a variable complexity. These are chemically not well-defined substances but rather a group of substances with the ability to bind proteins in aqueous solution. Their multiple phenolic hydroxyl groups lead to the formation of complexes primarily with proteins and to a lesser extent with metal ions, amino acids and polysaccharides. Tannins are tentatively classified into two classes: hydrolysable and condensed tannins (although tannins are known which have components of both hydrolysable and condensed tannins), and are considered to have both adverse and beneficial effects depending on their concentration and nature besides other factors such as animal species, physiological state of the animal and composition of the diet.

2.2. Preparation of plant extract

The general problem in the study of natural plant products is that their nature and amount are dependent on various factors, which must be controlled as far as possible. Some of these factors are: i) stress; the metabolic state of the plant may change when it is stressed in any manner. This can be a problem before as well as after harvesting a plant part for analysis. As cells die (the senescent process), the cellular integrity is lost and as a result the enzymes come in contact with substrates to which they are not normally exposed in living cells. In addition, it also increases the oxidation process, which is a problem with phenolics since these are prone to oxidation. On oxidation, phenolics oxidise to quinones and then polymerisation reaction could follow. If a plant is cut and dried under 'near ambient' conditions, which generally requires a large time to dry, the nature and content of phenolic compounds can change. In order to avoid these changes, the metabolic activities of the cells need to be curbed immediately. The next important step is to bring the chemical constituents into solutions for their measurement.

2.2.1. Collection, drying and storage of plant material

Leaf age and stage of development affect levels and nature of phenolics. Therefore, it is important to define the stage of maturity of plant and leaf as close as possible before collecting leaf material for analysis. When the collection site is close to the laboratory, the material can be transported to the laboratory in fresh state. The fresh material should be kept on ice and transported under dark conditions. Transportation of large amount of leaves in plastic bags should be avoided, since temperature in the bag could rise leading to sweating and wilting which can change the nature and level of phenolics. If liquid nitrogen is available, the better option is to freeze the sample and then freeze-dry the material without thawing it. Thawing can rupture cell membranes leading to changes in phenolics. If the material is frozen using a freezer, make sure that the material is not thawed during transport. Solid carbon dioxide should be used to transport such material. Once the material is dried, it should be kept in a dry place (preferably in a desiccator) in the dark. The freeze-dried material generally is hygroscopic. Light is also known to change the nature of phenolics. After freeze-drying, the cell structure is broken and the enzymes are in the native state. With the absorption of water, enzymes and phenolics can react, which can produce drastic changes in phenolics. The freeze-drying, though considered to be one of the safest method for preservation of phenolics, can lead to drastic changes if the storage conditions are not appropriate.

If a lyophilizer is not available, the plant material has to be dried under far from ideal conditions. The sample can be dried at about 50–52°C using a forced air oven. This will hasten the process of drying, and the enzymes present in the plant sample will not have much time to react with phenolics. Drying at temperatures higher than 55°C should be avoided, since it can lead to inactivation of phenolics or could decrease their extractability in solvents and affect the quantification.

The suggested procedures for collection, storage and drying of samples are given below.

2.2.2. Grinding of sample

Fresh or frozen materials are difficult to work with. Grinding could be a problem using these materials. Fresh material, when frozen using liquid nitrogen, can be ground using 'Polytron' homogenizers. One has to be cautious that the temperature does not rise during homogenization; increase in temperature can lead to enzymatic changes in phenolics. Phenolics are generally extracted in aqueous organic solvents. The moisture present in the fresh material needs to be taken into account while preparing organic solvents for extraction. An easier approach is to grind the dry sample. See below for the suggested procedure.

2.2.3. Extraction of tannins

The aim is to quantitatively diffuse phenolics present in the plant material to liquid phase. For the extraction process, a suitable solvent is required. Generally, aqueous methanol (50%) and aqueous acetone (70%) are popular choices. The latter has been reported by various workers to be better in extracting phenolics from tree leaves. One can try both these solvents for extraction, and then based on the efficiency of extraction of phenolics (using Folin-Ciocalteu method) and/or condensed tannins (using butanol-HCl method), one can decide the solvent to be used for a particular plant material.

The suggested procedure is given below.

2.2.4. Suggested procedures

Keeping in mind the above and the constraints and facilities in developing countries, following procedures emerged through discussions at the first Research Coordination Meeting held in Vienna in March 1999. These procedures were followed in the CRP by all the participating groups.

- *Collection of plant material.* Define stage of maturity of leaf and the plant as closely as possible. Hard twigs should not be taken for the studies.
- *Drying of plant material.* Since the amount of plant material required for *in vivo* studies is large, the material should be dried under shade. In places of high humidity, use of fans is advised. The sample should be stored only after it has a dry matter content of > 90%. A representative sample (taken using the established protocols) should be kept for tannin analysis. Representative samples should be taken just before storage and then again before feeding to livestock, and both sets of samples should be analysed. If possible, a representative sample of the fresh leaf material should also be frozen soon after collection and freeze-dried.
- *Storage of plant material.* The sample should be stored in a cool, dark and dry place.
- *Grinding of plant material.* About 500 g of the plant material should be ground first to pass a 2 mm screen. All the ground material including those parts remaining inside the mill should be taken, mixed well and approximately 100 g of this sample is again ground to pass through a 0.5 mm screen. This sample should be taken for tannin analysis. Take care that at any stage of the grinding, the sample temperature does not rise above 40°C. For fibre analysis, and *in situ* nylon bag and *in-vitro* gas techniques, the sample passed through 1 mm screen should be taken.
- *Extraction of tannins.* Use 70% aqueous acetone (v/v) for all assays except protein precipitable phenolic and protein precipitation assays including the radio-labelled protein precipitation assay; and for these two assays 50% aqueous methanol (v/v) should be used, since acetone interferes in the process of the complex formation. An ultrasonic water bath should be used for extraction of tannins (20 min) in these organic solvents as described below.
- Dried (finely ground; passed through a 0.5 mm screen) plant material (200 mg) is taken in a glass beaker of approximately 25 ml capacity. Ten ml of aqueous acetone (70%) is added and the beaker is suspended in an ultrasonic water bath (Branson 3210) and subjected to ultrasonic treatment for 20 min (2 x 10 min with 5 min break in between) at room temperature. The contents of the beaker is then transferred to centrifuge tubes and subjected to centrifugation for 10

min at approximately 3,000 *g* at 4°C (if refrigerated centrifuge is not available, cool the contents by keeping the centrifuge tube on ice and then centrifuge at 3,000 *g* using an ordinary clinical centrifuge). Collect the supernatant and keep it on ice. Transfer the pellet left in the centrifuge tube to the beaker using two portion of 5 ml each of 70% aqueous acetone and again subject the contents to ultrasonic treatment for 20 min. Centrifuge and collect supernatant as described above. Folin-Ciocalteu method is used for determination of total phenols in the two supernatants. In most of the studies, when the particle size of the ground plant material was very fine, the recovery of total phenols in the second supernatant was found to be < 5% of that in the first supernatant. Therefore, the second extraction step can be omitted. One can follow this approach and decide to extract the sample once or twice.

Notes:

Very long extraction at too high a temperature may lead to degradation and loss of phenolics.

Freshly prepared extract should be used for tannin analysis.

Tubes/container containing the extract should be kept on ice till the analysis is complete.

Pigments and fat can be removed from the dried leaf sample by extracting with diethyl ether containing 1% acetic acid before extracting tannins. The extraction can be hastened using an ultrasonic water bath (2 x 5 min) in a manner similar for tannin extraction. The time for pigment removal or tannin extraction using an ultrasonic water bath may vary from sample to sample.

2.3. Tannin assays

Use of proper analytical techniques for measurement of tannins is vital. Methods for quantification of tannins may be based on the chemical properties of tannins or their capability to bind substrates, particularly proteins. The methods described in this manual can be categorized as follows:

2.3.1. Chemical methods

(a) *Determination of total phenolics*. The method is based on the fact that phenolics are reducing agents. It may be noted that all tannins are phenolics but not all phenolics are tannins.

(b) *Determination of total tannins*. It is partly chemical, based on reducing property of tannins, and partly physical because tannins are measured as the reduction in phenolics that occur when a binding agent (polyvinyl polypyrrolidone, PVPP) is added to the extract. This method can not identify specific types of tannins in a mixture, i.e. information on presence or absence of condensed or hydrolysable tannins can not be obtained by this method.

(c) *Determination of condensed tannins (proanthocyanidins).* It is based on oxidative depolymerization of condensed tannins in butanol-HCl reagent. The presence of iron is considered to increase the reproducibility and sensitivity of the assay.

(d) *Determination of hydrolysable tannins.* The first method is based on hydrolysis of gallotannins to gallic acid under acidic conditions and measurement of the released gallic acid by reaction with rhodanine or by HPLC. In the second method reported here, hydrolysable tannins are converted to a common structural analyte, methyl gallate via methanolysis. The methyl gallate is reacted with potassium iodate to give a chromophore, the absorbance of which is measured at 525 nm.

Each method measures different types of tannins based on chemistry of the reaction between tannins/phenols and reagents used.

2.3.2. Protein precipitation/binding methods

(a) *Determination of protein precipitable phenolics and protein precipitation capacity.* These methods are based on the formation of tannin-protein complexes (tannins in the plant extract and the protein, bovine serum albumin). Tannins present in the complex are determined using ferric chloride assay for total phenolics. Iron forms a complex with phenols to give a pink chromatophore which is measured spectrophotometrically. In the method reported in this manual, the protein in the complex is measured by application of the dissolved tannin-protein complex on a cellulose acetate membrane, staining the protein with amido black, followed by image analysis to convert the stained protein to pixels. This method is a modification of the 'Filter paper-protein Ponceau S dye assay' which is described below.

(b) *Filter paper-protein Ponceau S dye assay.* The protein-tannin complexes are formed on a sheet of filter paper, and the protein bound to the complex is measured by dyeing it with Ponceau S dye. This dye is specific for proteins and does not bind tannins. The colour of the dye bound to proteins is eluted and measured spectrophotometrically.

(c) *Radial diffusion assay.* In this method, tannin molecules migrate through agarose gel which is impregnated with the protein, bovine serum albumin (BSA). The tannin-protein complex is formed in the gel which appears as a ring. The diameter of the ring is a measure of protein precipitation/binding capacity of tannins.

(d) *Radiolabelled bovine serumn albumin method.* The principle of this method is the same to that of those mentioned in (a) above, except that the ^{125}I- BSA is used to complex with tannins. The protein bound is calculated from the counts measured using a gamma counter. This is the most sensitive method for measurement of protein precipitated by a tannin.

2.3.3. *Tannin bioassay*

The feeds under investigation (or tannins) are incubated in absence and presence of a tannin-complexing agent, polyethylene glycol (PEG) of MW 4,000 or 6,000 in the *in vitro* rumen fermentation system containing rumen microbes. The affinity of PEG for binding to tannins is very high; it even breaks the already formed tannin-protein complexes and releases proteins from the complex. In the *in vitro* rumen fermentation system, release of gas and production of microbial mass are measured. The difference between these parameters observed in absence and presence of PEG is a measure of tannin activity in relation to their effects on rumen fermentation.

3. MEASUREMENT OF TOTAL PHENOLICS AND TANNINS USING FOLIN-CIOCALTEU METHOD

(According to Makkar *et al.* [1])

The method for total phenol is useful in order to know the efficiency of extraction of phenolics in solvents. This method can be coupled with the use of insoluble matrix, polyvinyl polypyrrolidone (PVPP; binds tannin-phenolics) for measurement of tannins (see below). The results can be expressed as tannic acid equivalent. The nature of tannic acid varies from one commercial source to the other [2]. Tannic acid from Merck was found to be most appropriate; highest efficiency of binding to bovine serum albumin (BSA), the least threshold at which it starts precipitating BSA and also the highest change in absorption per unit change in tannic acid level. Furthermore, there does not seem to be batch-to-batch variation in tannic acid from Merck.

3.1. Reagents

(1) *Folin-Ciocalteu reagent (1 N):* Dilute commercially available Folin-Ciocalteu reagent (2 N) with an equal volume of distilled water. Transfer it in a brown bottle and store in a refrigerator (4°C). It should be golden in colour. Do not use it if it turns olive green.

(2) *Sodium carbonate (20%):* Weigh 40 g sodium carbonate (x10 H_2O), dissolve it in about 150 ml distilled water and make up to 200 ml with distilled water.

(3) *Insoluble polyvinyl pyrrolidone (polyvinyl polypyrrolidone, PVPP):* This is commercially available from Sigma (P 6755).

(4) *Standard tannic acid solution (0.1 mg/ml):* Dissolve 25 mg tannic acid (TA) obtained from Merck in 25 ml distilled water and then dilute 1:10 in distilled water (always use a freshly prepared solution).

3.2. Preparation of calibration curve

Table IV. Preparation of calibration curve

Tube	Tannic acid solution (0.1 mg/ml) (ml)	Distilled water (ml)	Folin reagent (ml)	Sodium carbonate solution (ml)	Absorbance at 725 nm	Tannic acid (μg)
Blank	0.00	0.50	0.25	1.25	0.000	0
T1	0.02	0.48	0.25	1.25	0.112	2
T2	0.04	0.46	0.25	1.25	0.218	4
T3	0.06	0.44	0.25	1.25	0.327	6
T4	0.08	0.42	0.25	1.25	0.432	8
T5	0.10	0.40	0.25	1.25	0.538	10

3.3. Analysis of total phenols

Prepare tannin extract from the plant sample by following the method given in Section 2.2.4.

Take suitable aliquots of the tannin-containing extract (initially try 0.02, 0.05 and 0.1 ml) in test tubes, make up the volume to 0.5 ml with distilled water, add 0.25 ml of the Folin-Ciocalteu reagent and then 1.25 ml of the sodium carbonate solution. Vortex the tubes and record absorbance at 725 nm after 40 min. Calculate the amount of total phenols as tannic acid equivalent from the above calibration curve (Table IV). Express total phenolic content on a dry matter basis (x%).

Example:

50 μl tannin-containing extract in the assay mixture gives 0.531 absorption
= 9.896 μg tannic acid (TA) equivalent (from the standard curve).
Therefore, 1 ml extract has 9.896 / 0.05 = 197.9 μg TA = 0.198 mg TA.

200 mg leaf sample was extracted in 10 ml solvent.
Therefore, 100 mg leaf has 0.198 x 5 = 0.99 mg TA or 100 g leaf has 0.99 g TA.

If leaf contains 95% dry matter (DM),
TA in DM = 0.99 / 0.95 = 1.04%.

Call this x in subsequent calculations.

3.4. Removal of tannin from the tannin-containing extract

PVPP binds tannins. Weigh 100 mg PVPP in a 100 x 12 mm test tube. Add to it 1.0 ml distilled water and then 1.0 ml of the tannin-containing extract (100 mg

PVPP is sufficient to bind 2 mg of total phenols; if total phenolic content of feed is more than 10% on a dry matter basis, dilute the extract appropriately). Vortex it. Keep the tube at 4°C for 15 min, vortex it again, then centrifuge (3,000 *g* for 10 min) and collect the supernatant. This supernatant has only simple phenolics other than tannins (the tannins would have been precipitated along with the PVPP; the procedure for binding of tannins to PVPP is presently being modified, and the modification is to bind tannins to PVPP at pH 3 since PVPP binds maximally to tannins at this pH [3]). Measure the phenolic content of the supernatant as mentioned above (take at least double the volume (preferably three times) you used for total phenol estimation, because you have already diluted the extract two-fold and expect to lose tannin-phenols though binding with PVPP). Express the content of non-tannin phenols on a dry matter basis (y%).

Example, continued:

100 µl of the supernatant after PVPP treatment in the assay mixture gives 0.312 absorption
= 5.75 µg tannic acid (TA) equivalent (from the standard curve).

Therefore, 1 ml supernatant = 5.75 / 0.1 = 57.54 µg TA = 0.058 mg TA.

10 mg leaf has 0.058 mg TA (since the extract is diluted 2-fold during the test)

Therefore, 100 mg leaf sample has 0.058 x 10 = 0.58 mg TA

y = 0.58 / 0.95 (since dry matter of feed or seed = 95%)
y = 0.61%

But total phenolics in dry matter, x = 1.04%

(x–y) is the percentage of tannins as tannic acid equivalent on a dry matter basis.

Tannins (as tannic acid equivalent) = 1.04 - 0.61 = 0.43% in the dry matter.

Note: Ascorbic acid (generally added to prevent oxidation of phenols) interferes in this assay, and therefore should not be added to the solvent used for extraction of phenols/tannins

4. DETERMINATION OF CONDENSED TANNINS (PROANTHOCYANIDINS)

(According to Porter *et al.* [4])

4.1. Reagents

(1) *Butanol-HCl reagent (butanol-HCI 95:5 v/v):* Mix 950 ml n-butanol with 50 ml concentrated HCl (37%).

(2) *Ferric reagent (2% ferric ammonium sulfate in* 2 N *HCl):* Make 16.6 ml of concentrated HCl up to 100 ml with distilled water to make 2 N HCl. Dissolve 2.0 g ferric ammonium sulfate in this volume of 2 N HCl. This reagent should be stored in a dark bottle.

4.2. Analysis

Extract tannins from the plant sample as given in Section 2.2.4.

In a 100 mm x 12 mm glass test tube, pipette 0.50 ml of the tannin extract diluted with 70% acetone. The quantity of acetone should be large enough to prevent the absorbance (550 nm) in the assay from exceeding 0.6. It will depend on the quantity of condensed tannin expected in the sample, and occasionally will need to be determined by trial and error. To the tubes add 3.0 ml of the butanol-HCl reagent and 0.1 ml of the ferric reagent. Vortex the tubes. Cover the mouth of each tube with a glass marble and put the tubes in a heating block adjusted at 97 to 100°C (or in a boiling water bath) for 60 min. Cool the tubes and record absorbance at 550 nm. Subtract a suitable blank, which is usually the absorbance of the unheated mixture. If the extract has flavan-4-ols, a pink colour develops without heating. If this happens, use one heated blank for each sample, comprising 0.5 ml of the extract, 3 ml of butanol and 0.1 ml of the ferric reagent. Condensed tannins (% in dry matter) as leucocyanidin equivalent is calculated by the formula:

(A 550 nm x 78.26 x Dilution factor*) / (% dry matter)

This formula assumes that the effective $E^{1\%,\ 1\ cm,\ 550\ nm}$ of leucocyanidin is 460 [4].

*The dilution factor is equal to 1 if no 70% acetone was added and the extract was made from 200 mg sample in 10 ml solvent. Where 70% acetone is added (for example to prevent the absorbance from exceeding 0.6) the dilution factor is:

0.5 ml / (volume of extract taken)

Notes:

The presence of pigments may interfere in this method. The pigments can be removed by extracting the dried leaves with petroleum ether containing 1% acetic acid.

Vanillin-HCI is also used for determination of condensed tannins but this method is not specific. It measures condensed tannins as well as simple flavonoids. This method also has several other disadvantages. Readers are referred to Refs. [5, 6].

Ascorbic acid (generally added to prevent oxidation of phenols) does not interfere in this assay.

5. GALLOTANNIN DETERMINATION BY RHODANINE ASSAY

(According to Inoue and Hagerman [7])

The plant material was extracted as described in Section 2.2.4. The supernatant is designated as **supernatant A**.

5.1. Reagents

(1) *Rhodanine solution (0.667% w/v in methanol):* Weigh 667 mg rhodanine (Sigma, R 4375) and dissolve it in 100 ml methanol. Stable for at least 2 weeks when stored in a refrigerator.

(2) *KOH (0.5 N):* Dissolve 2.8 g potassium hydroxide in 100 ml distilled water.

(3) *Sulphuric acid solutions:* Prepare 0.3, 0.4, 0.6, 22 and 26 N sulphuric acid solutions by appropriately diluting 98% commercially available sulphuric acid (36 N).

5.2. Determination of free gallic acid

Pipette 200 µl **supernatant A** in a 160 mm x 12 mm culture test tube (4 tubes per sample). Remove acetone from the sample using a vacuum drying oven adjusted at 40°C (pressure 300 mbar) for about 2 h or by flushing with nitrogen gas (bring nitrogen gas beam as close as possible to liquid surface). Using nitrogen gas, it takes about 8–10 min to dry the sample. One can construct a multi-tube (8–10 outlets) assembly for flushing nitrogen gas, which allows simultaneous drying of 8–10 samples.

Add 200 µl of 0.2 N sulphuric acid to the tubes containing dried **supernatant A**. To three tubes add 300 µl of the rhodanine solution and to the fourth tube 300 µl methanol. This fourth tube acts as a blank. After 5 min add 200 µl of 0.5 N potassium hydroxide solution to all the tubes. Wait for 2.5 min and then add 4.3 ml of distilled water. After 15 min measure absorbance at 520 nm against the blank.

5.3. Determination of gallic acid present in free and in gallotannin forms

Pipette 3.34 ml of **supernatant A**, in duplicate, in the above mentioned culture test tube. Remove acetone by flushing the tubes with nitrogen gas. After removal of acetone, 1 ml of the supernatant remains (or make to 1 ml with distilled water). To it add 0.1 ml of 22 N sulphuric acid so that the final sulphuric acid concentration is 2 N. Freeze the contents and remove air from these culture tubes by using a vacuum pump. A special assembly can be constructed for removal of air. This assembly allows simultaneous removal of air and tightening of the cap of the culture tube. Keep these tubes at 100°C for 26 h to hydrolyse gallotannins to gallic acid. After hydrolysis, make up the volume to 11 ml by adding 9.9 ml distilled water. Sulphuric acid concentration in this solution is 0.2 N. This solution is addressed as '**hydrolysed supernatant A**'.

Pipette 200 μl of the **hydrolysed supernatant A** (4 tubes per sample; 1 blank and the rest 'test' samples). Now add 300 μl of the rhodanine solution to the 'test' tubes and 300 μl methanol to the blank tube. Wait for 5 min, and then add 200 μl of 0.5 N KOH solution. Again wait for 2.5 min and add 4.3 ml distilled water. After 10 min, measure absorbance at 520 nm.

Convert absorbance at 520 nm using the calibration curve (Table V).

5.4. Calibration curve

Stock gallic acid solution (1 mg/ml in 0.2 N sulphuric acid): Weigh 100 mg gallic acid and dissolve in approximately 80 ml of 0.2 N sulphuric acid and then make up the volume to 100 ml with 0.2 N sulphuric acid. It can be kept frozen for at least one month.

Working gallic acid solution (0.1 mg/ml): Dilute the gallic acid stock solution 1:10 with 0.2 N sulphuric acid (1 ml of the stock solution plus 9 ml distilled water). It can be stored in a refrigerator for at least two weeks.

Note: Vortex stir these solutions well.

Table V. Calibration curve

Gallic acid (μg)	Working solution (μl)	0.2 N Sulphuric acid (μl)	Rhodanine (μl)	Wait for at least 5 min	0.5 N KOH (μl)	Wait for at least 2.5 min	Distilled water (ml)
4	40	160	300		200		4.3
8	80	120	300		200		4.3
12	120	80	300		200		4.3
16	160	40	300		200		4.3
20	200	0	300		200		4.3
Blank	0	200	300		200		4.3

Wait for approximately 10 min, then measure absorbance at 520 nm.

Total gallic acid minus free gallic acid = gallotannins as gallic acid equivalent.

Note: Ascorbic acid (generally added to prevent oxidation of phenols) interferes in this assay, and therefore should not be added to the solvent used for extraction of phenols/tannins.

6. GALLOTANNIN DETERMINATION USING HPLC

(According to Makkar and Becker [8])

6.1. Preparation of sample for determination of free gallic acid

Pipette 1 ml of **supernatant A** (see Section 5) into the above mentioned culture tubes, remove acetone in a vacuum oven adjusted at 40°C (pressure 300 mbar). It takes about 3 h to remove acetone. Acetone can also be removed by flushing with nitrogen gas. Dry the contents completely in a heating block at 40°C by flushing with nitrogen. It takes about 25 min to dry the contents completely. It can be stored under nitrogen in a freezer.

Just before the analysis, add 750 µl distilled water to the residue, and dissolve the contents using an ultrasonic water bath (keep the tubes for about 5 min). Filter the contents through a 0.45 µm membrane filter (cellulose acetate or cellulose nitrate) before loading to the HPLC.

6.2. Preparation of sample for determination of gallic acid present in free and in gallotannin forms

Pipette 2 ml of the **hydrolysed supernatant A** (see above) into a 25 to 50 ml capacity tube/beaker and then add 2 ml buffer (2.304 g $NH_4H_2PO_4$ dissolved in 1000 ml distilled water) to it. Adjust the pH between 6.3 and 6.8 using 8 M KOH solution (approximately 120 µl will be required), and record the exact amount required for the pH adjustment. Do not exceed pH 7.0 because the sample will immediately get oxidized. In case the pH rises beyond 7, discard it, and start afresh with 2 ml of the hydrolysed supernatant. Freeze overnight, thaw it and then filter through 0.45 µm membrane filter before loading to the HPLC.

HPLC conditions and analysis:

Column: 250 mm x 4.6 mm filled with Nucleosil 120-5 C18 (Macherey Nagel GmbH & Co. KG, Postfach 10 13 52, D- 52313 Duren).

Flow rate: 1.2 ml/min.

Column temperature: Ambient (ca 22°C).

Injection volume: 20 µl.

Gallic acid stock solution: Dissolve 400 mg gallic acid in 1000 ml distilled water.

Gallic acid working solution (6 mg/1000 ml): Pipette 150 μl of the stock solution and make the volume to 10 ml with distilled water. Linearity between area and concentration was observed in the range of 2 to 10 mg gallic acid/ml on injection of 20 μl of gallic acid solution).

Elution time of gallic acid: Between 14 and 15.5 min.

Solvents: Buffer A; H_2O – methanol – H_3PO_4 (975.5:19.5:1; v/v/v), and Buffer B; methanol – H_2O (700:300; v/v) used as a gradient elution (Table VI).

Table VI: Gradient used

Time (min)	Solvent A (%)	Solvent B (%)
0.0	100	0
15.0	100	0
22.0	0	100
25.0	0	100
30.0	100	0
33.0	100	0

Detection wavelength: 280 nm.

Note: Ascorbic acid does not interfere in this assay, and therefore can be added to the solvent used for extraction of phenols/tannins.

7. DETERMINATION OF HYDROLYSABLE TANNINS (GALLOTANNINS AND ELLAGITANNINS) AFTER REACTION WITH POTASSIUM IODATE

(According to Hartzfeld *et al.* [9]).

7.1. Reagents

(1) *Anhydrous methanol, reagent or analytical grade.*

(2) *Concentrated sulphuric acid (18 M):* 98% commercially available sulphuric acid is 18 M.

(3) *Ethanolamine* (a liquid, commercial preparation is 100% ethanolamine).

(4) *Ammonium acetate (3.7 M):* Weigh 281.3 g ammonium acetate, dissolve it in approx. 800 ml distilled water and then make it up to 1 liter with distilled water. Ammonium acetate comes as a hygroscopic crystal, so after the bottle has been opened the crystals may be quite wet. Approximate the mass needed as accurately as possible, and keep the container tightly closed.

(5) *HCl (0.3 M):* Dilute 10 ml HCl (37%) to 400 ml with distilled water.

7.2. Procedure

7.2.1. Methanolysis and pH adjustment.

Weigh approximately 20 mg samples of dried, ground plant sample into 16 x 125 mm (20 ml) pyrex screw-top tubes with Teflon cap liners, and record the exact mass of each sample (the top part of tube acts as a condenser during the methanolysis, so do not substitute smaller tubes). Add to each tube 2 ml methanol and 200 µl of concentrated sulphuric acid (18 M), and cap the tubes finger tight so that solvent does not evaporate during heating. Place the capped tubes in a heating block previously brought to 85°C and allow to react for 20 h at 85°C. Centrifuge samples in a table top centrifuge (3,000 *g,* 10 min). Quantitatively transfer the supernatant to 5 ml graduated cylinder by washing the methanolysis tube three times with a minimum volumes of distilled water, and recentrifuging as necessary. Adjust the volume to 3.0 ml with distilled water. Add four 50 µl aliquots of ethanolamine with gentle swirling between each addition (the heat of neutralization is high so the ethanolamine must be added in these small aliquots). To buffer the sample, then add 500 µl of 3.7 M ammonium acetate and adjust the pH to 5.5 $\pm$ 0.1 using a pH meter and small volumes of dilute ethanolamine or dilute sulphuric acid (this must be done carefully, as the final pH and volume is critical to success of the method). Bring the sample to a final volume of 4 ml with distilled water, and mix thoroughly. Samples prepared this way can be stored tightly capped at 4°C for up to 48 h. A precipitate may form during storage; in that case recentrifuge briefly the solution to remove the precipitate before analysis.

7.2.2. Standard and controls

The method is standardized with methyl gallate. To control for any losses during methanolysis, 2 ml of 5 mg/ml methyl gallate dissolved in methanol is methanolysed and pH adjusted as described above for the plant sample. A phenol-free, pH 5.5, blank reagent is prepared by treating a mixture of 2 ml methanol and 200 μl concentrated sulphuric acid, exactly as described above for the sample. This blank reagent does not contain any plant sample or standard polyphenolic compound, and is also used to dilute the standard when preparing a standard curve, or the methanolysed and pH-adjusted sample solution before reacting with potassium iodate.

7.2.3. Reaction with potassium iodate

The pH-adjusted (5.5 $\pm$ 0.1) samples, obtained as described above, are analysed with potassium iodate. The reaction yields a characteristic pigment with λ_{max} at 525 nm. Dispense up to 100 μl of sample (methanolysed and pH-adjusted sample comprised of plant samples or standard methyl gallate) into a 2 ml microcentrifuge tube. Add the pH 5.5 blank reagent, as necessary, to bring the sample volume to 100 μl (for methanolysis products of most plants, 100 μl aliquot of the methanolysis solution; and for standards, zero to 100 μl of the pH-adjusted methyl gallate solution are generally used). Add 350 μl distilled water and 1 ml methanol to each tube and then vortex. Cap the tubes tightly and place in a 30°C water bath.

Some flavanoids and other plant constituents react under the conditions of the assay to form brownish pigments that absorb weakly at 525 nm. To correct the background colour, prepare a background mixture for each sample exactly like the reaction mixture except that the 350 μl of water is replaced by 350 μl of 0.3 M HCl.

At timed intervals, add to each tube 40 μl of 5% (w/v) potassium iodate solution. Cap the tubes, vortex them and return to a 30°C water bath. Exactly after 50 min of adding the potassium iodate solution, read the absorbance at 525 nm against water. The absorbance of the background mixture (against water) is subtracted from that of the corresponding sample.

Under the conditions of the assay, the linear relationship between methyl gallate and absorbance is obtained at least up to 120 μg. The limit of detection by this method is 1.5 μg methyl gallate. No interference has been recorded on using this method to determine methyl gallate after methanolysis of purified hydrolysable tannins such as tannic acid, epigallocatechin gallate, and oenothein.

The results are expressed as methyl gallate equivalent.

Notes:

This functional group method for determination of gallic acid esters was originally described by Haslam [10]. The method was modified slightly by Bate-Smith [11] and more recently was examined by Willis and Allen [12]. The well known limitations of the method are the consequences of the instability of the red oxidation products of the reaction between gallic acid esters and potassium iodate; different color yields for structurally distinct hydrolysable tannins; and the poor understanding of the dependence of product accumulation with respect to reaction solvent, pH and temperature.

The method reported in this manual [9] is a modification of the original method of Haslam [10] and Bate-Smith [11]. In this modified method hydrolysable tannins are converted to a common structural analyte, methyl gallate via methanolysis. The methyl gallate is reacted with potassium iodate. Another common analyte of hydrolysable tannins could be gallic acid, but gallic acid is oxidized by potassium iodate to form a non-specific yellow compound that is not useful for selective determination.

In the modified method, conversion of hydrolysable tannins to methyl gallate has been optimized using HPLC to assess acid strength, temperature and time required to maximize yield of methyl gallate, which has enhanced the reproducibility of the potassium iodate method.

It is essential to use anhydrous methanol for methanolysis, because even a traces of water cause production of a mixture of gallic acid and methyl gallate.

8. DETERMINATION OF PROTEIN-PRECIPITABLE PHENOLICS

(According to Makkar *et al.* [13]).

8.1. Reagents

(1) *Acetate buffer (pH 4.8 to 4.9, 0.2 M):* Pipette 11.4 ml glacial acetic acid to about 800 ml distilled water. Adjust pH of this solution to 4.8 to 4.9 with 4 N sodium hydroxide solution, and bring the final volume to 1 litre. To it add 9.86 g NaCl to make its concentration 0.17 M.

(2) *Sodium dodecyl sulfate solution (SDS) (1% w/v):* Dissolve 1 g SDS in 100 ml of distilled water.

(3) *SDS-triethanolamine (TEA) (1% SDS (w/v) and 7% (v/v) triethanolamine in distilled water) solution:* To 7 ml of triethanolamine add 93 ml distilled water and dissolve 1 g SDS in this solution.

(4) *Ferric chloride reagent (0.01 M ferric chloride in 0.1 M HCl:.* For making 0.1 M HCI, dilute 4.2 ml concentrated HCl (37%) to 500 ml with distilled water. Dissolve 0.81 g ferric chloride in 500 ml of 0.1 M HCI. Filter and store the contents in a brown bottle.

(5) *Glacial acetic acid.*

(6) *BSA solution:* Dissolve 100 mg BSA (fraction V) in 100 ml of the acetate buffer.

8.2. Formation of the tannin-protein complex

Extract tannins from the plant sample by following the procedure presented in Section 2.2.4.

To 2 ml of the BSA solution (containing 1 mg BSA/ml acetate buffer), add 50% methanol and increasing levels of the tannin-containing extract to make 3 ml. For example, use 0.90, 0.85, 0.80, 0.75, 0.70, 0.65, 0.60, 0.50 ml of 50% methanol with 0.10, 0.15, 0.20, 0.25, 0.30, 0.35, 0.40, 0.50 ml of the extract (Table VII); this may vary depending on the amount of tannins in the sample. Vortex the contents. Allow the mixture to stand in a refrigerator (4°C) overnight. Centrifuge at about 3,000 *g* for about 10 min. Remove the supernatant carefully without disturbing the precipitate. Add 1.5 ml of 1% SDS solution to the precipitate and vortex it to dissolve the precipitate.

Note: It is essential to use 50% aqueous methanol for extraction of tannins from the plant material since acetone interferes in the protein-precipitation assay

8.3. Determination of tannins (phenolics) in tannin-protein complex

Take an aliquot (1 ml) of the above dissolved complex. Add 3 ml of SDS-triethanolamine solution. Then add a 1-ml portion of the ferric chloride reagent. Record absorbance at 510 nm after 15–30 min[a]. Convert the absorbance to tannic acid equivalent, using a standard curve (Table VIII). Multiply the values obtained by 1.5 to obtain tannins in the complex. Draw a linear regression between tannins precipitated as tannic acid equivalent and mg leaf (in aliquot taken for the assay). The slope of the curve (mg tannic acid precipitated/mg leaf; let it be x) represents the protein-precipitable phenolics in the sample.

Table VII. An example (*Acacia cyanophylla*)

Tube	Extract (μl)	Leaf (mg)*	Absorbance at 510nm	TA (mg)**	TA in complex (mg)***
1	100	2	0.121	0.054	0.081
2	150	3	0.167	0.077	0.116
3	200	4	0.234	0.109	0.164
4	250	5	0.292	0.136	0.204
5	300	6	0.341	0.160	0.240
6	350	7	0.422	0.199	0.299
7	400	8	0.472	0.222	0.333
8	500	10	0.591	0.280	0.420

TA, tannic acid (Merck)

* 200 mg leaf is extracted in 10 ml 50% aqueous methanol.

** Conversion of absorbance at 510 nm to mg tannic acid by the standard curve (see below).

*** Obtained by multiplying values in the previous column (which correspond to 1 ml of the soluble tannin-protein complex) by 1.5, because the tannin-protein complex is dissolved in 1.5 ml of 1% SDS.

Protein precipitable phenolics (x; mg tannic acid equivalents precipitated/mg leaf dry matter) for the above example = 0.043/0.953 = 0.045, since dry matter of the leaves was 95.3%.

[a] Adapted from Hagerman, A.E. and Butler, L.G. J. Agric. Food Chem. 26, 809 (1987).

Table VIII. Calibration curve for phenolic content

Tube	TA solution * (ml)	SDS, 1% (ml)	SDS-TEA (ml)	Ferric chloride (ml)	Absorbance at 510 nm	TA (mg)
Blank	0	1.0	3.0	1.0	0.000	0.00
T1	0.1	0.9	3.0	1.0	0.107	0.05
T2	0.2	0.8	3.0	1.0	0.225	0.10
T3	0.3	0.7	3.0	1.0	0.319	0.15
T4	0.4	0.6	3.0	1.0	0.426	0.20
T5	0.5	0.5	3.0	1.0	0.527	0.25

TA, tannic acid (Merck)

* TA solution: 0.5 mg/ml in 1% SDS.

Note: The protein in the dissolved complex can also be measured by alkali hydrolysis of protein to amino acids and measurement of the amino acid released by ninhydrin [14].

8.4. Protein precipitable phenolics as percentage of total phenolics

8.4.1. Determination of total phenolics in the original extract

Take different aliquots (generally 0.05, 0.10, 0.15, 0.2, 0.25, 0.30 ml, but this may vary depending on the amount of phenolics in the sample) of the extract (200 mg in 10 ml of 50% methanol; Section 2.2.4.), make up to 1 ml with 1% of SDS, and add 3 ml of the SDS-triethanolamine solution and 1 ml of the ferric chloride reagent. Record absorbance at 510 nm as described in Table IX. Convert the absorbance to tannic acid equivalent using the standard curve (Table VIII). Draw a linear regression between tannic acid equivalent and mg leaf (in the aliquot taken). The slope of the curve (mg tannic acid equivalent/mg leaf; let it be y) represents total phenolics.

Protein-precipitable phenolics have already been measured as x (Section 8.3).

The percentage of total phenolics which precipitate protein = (x / y) x 100.

Table IX. An example (Acacia cyanophylla)

Tube	Extract (μl)	Leaf* (mg)	Absorbance at 510 nm	TA**
1	50	1	0.145	0.066
2	100	2	0.280	0.131
3	150	3	0.404	0.190
4	200	4	0.532	0.251
5	250	5	0.674	0.319
6	300	6	0.824	0.391

TA, tannic acid (Merck)
* 200 mg leaf is extracted in 10 ml 50% aqueous methanol.
** Calculated from the calibration curve below.
Total phenolics (y; mg tannic acid equivalent/mg leaf dry matter)
= 0.064/0.9535 = 0.067, since dry matter of the leaves was 95.35%.

Therefore, protein precipitable phenolics as percentage of total phenolics
= (x / y) x 100 = (0.045 / 0.067) x 100 = 67.2.

8.5. Determination of proteins in tannin-protein complex using image analysis

(According to Hoffmann *et al.* [15]).

The modified dot blot assay allows the determination of protein concentration from the tannin-protein complex dissolved in 1% sodium dodecyl sulphate (SDS). The publication cited above works with a fluorochrome (benzoxanthene yellow), which requires special equipment for video documentation. If this is not available, the common protein stain amido black can be used instead. The protocol below refers to this simplified approach.

8.5.1. Sample preparation

Extract finely ground plant samples in 50% methanol (20 mg/ml) as described in Section 2.2.4 of this manual. Follow the protocol for the formation of the tannin-protein complex and its dissolution in 1% SDS as given in Section 8.2. Use these dissolved complexes directly in the dot blot assay.

8.5.2. Modified dot blot assay (amido black)

This assay works in the presence of SDS, but due care has to be taken to maintain a constant SDS concentration in all samples and in the standard used for calibration.

8.5.3. *Calibration*

Table X. Calibration curve for protein

µl BSA stock solution (2 mg/ml in H_2O)	µl H_2O	µl 2% SDS	Final protein conc. (mg/ml)
20	180	200	0.1
40	160	200	0.2
80	120	200	0.4
120	80	200	0.6
160	40	200	0.8
200	0	200	1.0

8.5.4. *Procedure*

Mark the upper left corner of a cellulose acetate membrane (Sartorius 12200) by cutting the edge for spatial orientation. Place the membrane in a holder (shown below) with an application grid and fix it by inserting two pins. The holder guarantees that, within the cavities the membrane does not touch the support and the samples can be quantitatively applied. The application pattern defines each sample by its position on the grid. A sketch of the holder is given below; it can be manufactured from any kind of inert plastic.

In the centre of the cavities, apply very slowly 2 µl of the standard or sample in triplicate. The protein concentration should be in the range of 0.1–1.0 mg/ml. Gilson/Eppendorf pipettes or 2 µl glass capillaries can be used for sample application (when the sample is completely absorbed by the membrane, there is still a small margin between the edge of the spot and the cavity wall). Let the membrane dry completely at room temperature and then remove it from the holder. Slowly immerse the membrane into the staining solution (0.5% amido black in methanol-acetic acid solution; 9: 1; v/v) and stain for 5 min with gentle agitation. Close the staining tray with a lid to avoid evaporation of methanol. Destain 3 x 5 minutes and 1 x 15 min in the methanol-acetic acid solution (9: 1; v/v). Again, keep the tray closed and never let the membrane dry during transfers (on evaporation of methanol the acetic acid will concentrate and destroy the membrane). Equilibrate the membrane for 2 min in the 6 : 3 : 1 (v/v/v) solution of butanol-methanol-acetic acid (as acetic acid evaporates faster than butanol, there is no more danger of destroying the membrane).

Record a digital picture of the wet membrane; avoid air bubbles between the membrane and the support. A video camera system or a flat-bed scanner may be used for taking the picture. Take grey-scale pictures only (quantification will not work with coloured pictures). Use an image analysis program (e.g., AIDA 2.31, Raytest GmbH, Straubenhardt, Germany) able to count pixels values in defined areas to convert the staining intensity of the spots to numbers. There will be a faint background, which should be subtracted for high accuracy. Depending on the image analysis program this can be done by defining a local background around each sample spot (see example given below), or by evaluating an empty area on the membrane, which is of equal size as the sample spots. The net pixel numbers per sample spot will then be converted to protein concentrations according to the calibration series blotted on the same membrane (Table X).

8.5.5. Membrane Holder

Two Polycarbonate plates (135 mm x 83 mm x 4 mm) are fixed on top of each other and holes of 8 mm diameter drilled through both plates. The distance between the centre points of the cavities is 10 mm, resulting in a 2 mm distance between neighbouring wells. In the example shown below, the grid has 7 x 12 positions. Due to a recent change made by the firm (Sartorius) in the membrane format, 6 x 12 positions are now sufficient. The smaller holes (1 mm diameter, 10 mm spacing) are used to insert two metal pins to fix the membrane between the plates for sample application; only two of them are needed here.

FIG.2. Photograph of the membrane holder

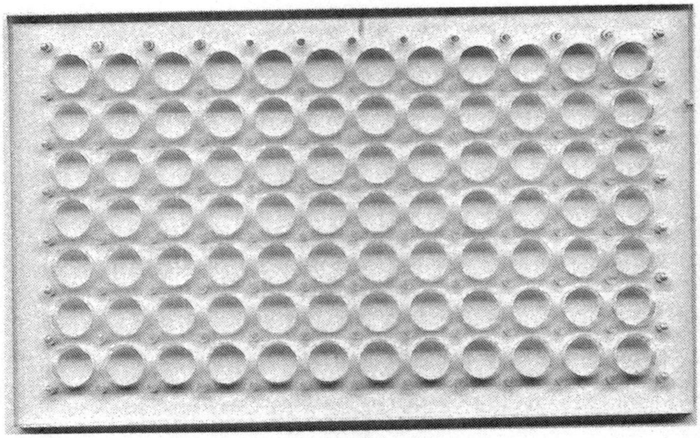

8.5.6. *Application pattern*

Descr.: –
Assay: –
Stain: –
Date: –

–	Sample	mg/ml	Comment	Sample	Dilut.	fraction	Sample	Dilut.	fraction	Sample	Dilut.	fraction
A	BSA	–		–	–	–	–	–	–	–	–	–
B	BSA	–		–	–	–	–	–	–	–	–	–
C	BSA	–		–	–	–	–	–	–	–	–	–
D	BSA	–		–	–	–	–	–	–	–	–	–
E	BSA	–		–	–	–	–	–	–	–	–	–
F	BSA	–		–	–	–	–	–	–	–	–	–

data (mg/ml)	1	2	3	4	5	6	7	8	9	10	11	12
A												
B												
C												
D												
E												
F												

8.5.7. *Stained membrane with regions defined for evaluation*

FIG.3. Photograph of the stained membrane

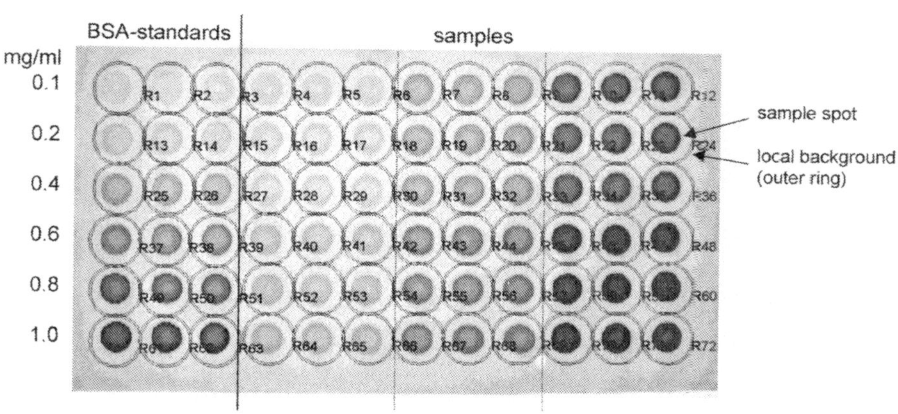

9. PROTEIN-BINDING CAPACITY BY FILTER PAPER ASSAY

(According to Dawra *et al.* [16]

9.1. Reagents

(1) *Tannic acid solution:* Dissolve 100 mg tannic acid (Merck, Germany) in 100 ml of 50% aqueous methanol. Add 100 mg ascorbic acid to it to minimize oxidation of tannic acid during handling.

(2) *Acetate buffer (pH 5, 0.05 M):* Pipette 2.85 ml glacial acetic acid to about 800 ml distilled water. Adjust pH of this solution to 5.0 with 4 N sodium hydroxide solution, and bring the final volume to 1 litre.

(3) *Bovine serum albumin (BSA) solution:* Dissolve 200 mg of bovine serum albumin (fraction V) in 100 ml of 0.05 M of the acetate buffer (pH 5, 0.05 M).

(4) *Dye solution:* Prepare 3% (w/v) trichloroacetic acid (TCA) solution in distilled water. Dissolve 0.2 g of Ponceau S dye in 100 ml of the TCA solution.

(5) *Acetic acid solution (0.2%, v/v):* Add 2 ml of glacial acetic acid to 998 ml distilled water.

(6) *Sodium hydroxide solution (0.1 N):* Weight 4 g sodium hydroxide and dissolve it in approximately 500 ml distilled water and then make up the volume to 1 litre with distilled water.

(7) *Acetic acid solution (10%, v/v):* Add 10 ml of glacial acetic acid to 990 ml distilled water.

9.2. Preparation of plant extract

As described above in 70% acetone (see Section 2.2.4.). Acetone does not interfere in this method.

9.3. Assay procedure

Take 1 MM Whatman paper chromatography sheet and cut it into an appropriate size (depends on the number of samples to be analysed). Draw squares of approximately 1.5–2 cm using a light lead pencil on the chromatography sheet.

Apply different aliquots (5–25 µl containing 5–25 µg tannic acid) on the sheet; each aliquot, at least in triplicate and on three different squares. Similarly, apply appropriate aliquots (10–50 µl; depending on the amount of tannins present) of the plant extract on the middle of squares on the chromatography sheet. Allow the spots to dry and spray immediately with the BSA solution until the paper is wet. After 30 min, wash the paper with the acetate buffer (pH 5, 0.05 M) with three 10-min changes with slight shaking to remove the unbound BSA. The paper was stained with 0.2% Ponceau S dye solution by keeping the strips dipped for 10 min in the stain solution. The staining solution should not be used in successive experiments. Wash the stained strips in 0.2% acetic acid solution until no more colour is eluted from the strips. Normally, this requires three washings to make the background clear. Air-dry the strips and cut the stained areas. To prepare a corresponding blank, stain simultaneously a chromatography sheet and wash in a manner similar to other samples. Cut the stained area into small pieces and elute the colour by adding 3 ml of 0.1 N sodium hydroxide solution and vortexing, followed by addition of 0.3 ml of 10% acetic acid and centrifugation at approximately 2,500 *g*. Measure the absorbance of the colour at 525 nm against corresponding blank. Convert this absorbance to protein content by using a standard curve. For preparing a standard curve, apply different concentrations of BSA (5–50 µg; 5–50 µl of 1 mg/ml BSA solution in the acetate buffer) as separate spots (at least in triplicate for each concentration) on a chromatography sheet and cut into strips. Stain these strips with the dye solution, wash and read the colour as described above for the samples.

Note: Ascorbic acid does not interfere in this assay, and therefore can be added to the solvent used for extraction of phenols/tannins.

10. RADIAL DIFFUSION ASSAY

(According to Hagerman [17])

10.1. Chemicals

Agarose: Type I, gel point 36°C (Sigma A-6013), bovine serum albumin (fraction V) (Sigma A-3350), glacial acetic acid, ascorbic acid and sodium hydroxide.

10.2. Reagents

(1) *Sodium hydroxide solution (4 N):* Weigh 8 g of sodium hydroxide and dissolve it in approximately 40 ml distilled water, make up the volume to 50 ml with distilled water.

(2) *Acetate buffer (pH 5.0, 0.05 M) containing 60 μM ascorbic acid:* Pipette 2.85 ml glacial acetic acid to about 800 ml distilled water and add to it 10.6 mg ascorbic acid. Adjust pH of this solution to 5.0 with 4 N sodium hydroxide solution, and bring the final volume to 1 litre.

10.3. Preparation of plates

Weigh 2.5 g agarose into 250 ml of the acetate buffer. Heat (boil for approx. 15 min) with continuous stirring on a magnetic stirrer until the agarose dissolves. Bring this solution to 45°C (takes approx. 5 min) by keeping the vessel containing this agarose solution into a water bath set at approximately 45°C. Add 250 mg BSA and dissolve it in the agarose solution without allowing the solution to cool. This solution will be sufficient for preparation of approximately 20 Petri dishes (internal diameter 8.5 cm). Use a 10 ml of glass pipette with a large tip opening or 10 ml of disposable syringes and dispense approximately 10 ml of the solution into each Petri plates kept on a flat surface. Take care not to introduce air bubble while transferring the agarose solution containing BSA. Make sure that the solution covers all the surface of Petri plates. Allow the solution to harden. Cover the Petri dishes and seal them with a strip of Parafilm in order to prevent drying and cracking of the agarose layer. These plates can be stored for two to three weeks in a refrigerator without losing sensitivity of the assay.

10.4. Assay procedure

On the day of performing the assay, take out the Petri dishes from the refrigerator, bring them to room temperature and then open them. Using a well puncher (4 mm) punch three to four wells, far apart, in the solidified agarose in

Petri dishes. Remove the plugs of agarose from the Petri dishes gently so that it comes along the puncher or remove the plug using a gentle suction.

Pipette 10 µl (using a micropipette) of the extract (Section 2.2.4.) in each well. If the preliminary experiment showed that tannin content is low, add 10 µl portion two or three times. The wells should not be allowed to completely dry between successive addition of aliquots. Cover the Petri dishes and again seal using Parafilm. Place the plates on a horizontal platform in an incubator adjusted at 30°C. After 96 h, remove the Petri dishes, uncover and measure the diameter of the ring formed. The amount of tannins is proportional to the square of the diameter. The square of the diameter of the ring formed can be calibrated by taking different concentration of tannic acid or any other standard tannin, and the results can be expressed equivalent to a standard.

Note: Acetone does not interfere in this method. The plant can be extracted in 70% aqueous acetone (see Section 2.2.4.)

11. TANNIN BIOASSAY

(According to Makkar *et al.* [3])

11.1. Sample preparation

Dried leaves should be passed through a 1 mm sieve.

11.2. Reagents

(1) *Bicarbonate buffer solution:* Dissolve 35 g sodium bicarbonate ($NaHCO_3$) and 4 g ammonium carbonate (NH_4HCO_3) in approximately 500 ml distilled water and then make up the volume to 1 litre with distilled water.

(2) *Macromineral solution:* Dissolve 6.2 g potassium dihydrogen phosphate (KH_2PO_4), 5.7 g disodium hydrogen phosphate (Na_2HPO_4), and 0.6 g magnesium sulphate ($MgSO_4.7H_2O$) in approximately 500 ml distilled water and then make up the volume to 1 litre with distilled water.

(3) *Micromineral solution:* Dissolve 10 g manganese chloride ($MnCl_2.4H_2O$), 13.2 g calcium chloride ($CaCl_2.2H_2O$), 1 g cobalt chloride ($CoCl_2.6H_2O$), 8 g ferric chloride ($FeCl_3.6H_2O$) in approximately 50 ml distilled water and then make up the volume to 100 ml with distilled water.

(4) *Resazurine:* Dissolve 0.1 g resazurine in 100 ml distilled water.

(5) *Reducing solution:* Dissolve 996 mg sodium sulphide ($Na_2S.9H_2O$) in 94 ml distilled water and then add 6 ml of 1 N sodium hydroxide solution (dissolve 4 g sodium hydroxide in 100 ml distilled water for 1 N sodium hydroxide).

11.3. Weighing of samples and preparation of syringes

Tear a specially made scoop (approximately 4 cm in length and 1 cm in depth/radius; standard sodium hydroxide-containing plastic container can be cut horizontally to half to form the scoop) on an analytical balance. Weigh 500 mg feed sample (passes through 1 mm sieve) in the scoop and then insert a 5 ml capacity pipette or a glass rod into the narrow end of the scoop and transfer the sample from the scoop into 100 ml calibrated glass syringes. Weigh 1 g tannin-complexing agent, polyethylene glycol, PEG (MW 4,000 or 6,000) on the scoop and transfer it also into syringes similar to those for the feed samples. The feedstuffs with and without the tannin-complexing agent are incubated at least in triplicate.

11.4. Preparation of *in vitro* rumen fermentation buffer solution

Collect the rumen fluid and particulate matter before the morning feed from two cattle fed on roughage diet, homogenize, strain and filter through four layers of cheese cloth. Keep all glassware at approximately 39°C and flush these with carbon dioxide before use. Carbon dioxide is heavier than air and hence it remains in the glassware for a reasonable period provided the container is not inverted upside down. The strained rumen fluid is kept at 39°C under carbon dioxide and should be prepared just before start of the incubation. As the amount of feed taken is 500 mg, composition of the medium is according to Tilley and Terry [18]. Menke *et al.* [19] reduced the rumen buffer volume per syringe by half as they used 200 mg of the substrate because of the limited volume of the syringes and the inconvenience of emptying the syringes. Here, besides recording the gas volume, we are interested to use the fermented material for various analyses; therefore, the amount of substrate taken is 500 mg. There is an inherent error associated with gravimetric determination of the fermented residue which (error) is large if 200 mg feed is taken in place of 500 mg.

➢ *Medium composition and incubation procedure*

Bicarbonate buffer solution	630.00 ml
Macromineral solution	315.00 ml
Micromineral solution	0.16 ml
Resazurine	1.60 ml
Distilled water	945.00 ml
Freshly prepared reducing solution	60.00 ml
Rumen fluid	660.00 ml
(see above for collection and preparation)	

The above volume is sufficient for 60 syringes (40 ml/syringe) plus 10% extra.

Mix, in the order mentioned above, all the above-mentioned solutions, except rumen fluid and the reducing solution, in a 3 or 5 litre capacity glass container. The container is kept in a water bath adjusted at 39°C. This water bath is a plastic rectangular container (400 cm x 300 cm x 200 cm) filled with water, the temperature of which is adjusted at 39°C using a portable thermostat suspended from the top of the plastic container in water. This plastic water bath is kept on a magnetic stirrer. The contents are flushed with carbon dioxide and kept stirred using a magnetic stirrer. After about 5 min, add the reducing solution and keep the mixture stirring and flushing with carbon dioxide at 39°C. When the mixture has been reduced (blue colour of the dye changes to pink and then to colourless; it takes about 15–20 min for the reduction process to complete and during this time we generally homogenized and strained the rumen liquor and the particulate material collected from cattle), add 660 ml of the rumen fluid. Keep this mixture stirring and flushing with carbon dioxide for another 10 min. Transfer a portion (40 ml) of the rumen-fluid medium into each syringe using a dispenser, and incubate in a water bath at 39°C. For filling 60 syringes, after some practice should take 30–35 min. After completion of the filling-up process, shake the syringes well and then transfer them to the water bath. Shake all the syringes every hour for the first four hours and then every two hours. Generally, the incubation is started at about 7.30 a.m. and after 12 h of the incubation, the syringes are not shaken till the termination of the incubation (16 h and/or 24 h). The gas volume is recorded after 2, 4, 6, 8, 10, 12, and 16 h or 24 h. The net gas production is calculated by subtracting values for the blank. The blanks (at least three in number) contain only the rumen-fluid medium and no feed material. Addition of PEG to blanks does not affect the gas production from blanks, suggesting that it is inert.

The difference between the net volume of gas produced from syringes with and without PEG is a measure of tannin effect. PEG binds tannins and inactivates them. From the gas volume, production of short-chain fatty acids (SCFA) can be calculated from the following equation [20].

Relationship between SCFA (mmol/syringe) and atmospheric pressure-corrected gas volume (the experiment was conducted in Hohenheim, Stuttgart which is at a height of 500 m from the mean see level). The equations are valid at 1 atmospheric pressure.

SCFA (mmol/40 ml) = -0.0601 + (0.0239 x Gas (ml)); n = 39 (without PEG), $R^2 = 0.953$; Standard error of estimate = 0.109; P<0.001.

SCFA (mmol/40 ml) = 0.0521 + (0.0207 x Gas (ml)); n = 37 (with PEG); missing observations = 2, $R^2 = 0.925$; Standard error of estimate = 0.108; P<0.001

SCFA (mmol/40 ml) = -0.00425 + (0.0222 x Gas (ml)); n = 76 (with and without PEG), $R^2 = 0.935$; Standard error of estimate = 0.115; P<0.001

These equations have been generated on incubation of leaves from trees and browses (39 in number).

One of the many factors which determines gas volume is the molar proportions of SCFA. 'Near-one' R^2 observed (see above) suggest that the total SCFA production can be calculated with reasonable accuracy from the volume of gas produced.

11.5. Measurement of microbial protein in the residue (apparent undigested residue) left after fermentation

(According to Makkar and Becker [21])

11.5.1. Preparation of apparent undigested residue

After 16 or 24 h of fermentation, centrifuge (20,000 *g*, 30 min, 4°C) the contents and discard the supernatant. Wash the pellet with distilled water followed by centrifugation (20,000 *g,* 30 min, 4°C). Lyophilize the pellet consisting of undigested substrate and microbial mass. The lyophilized residue has been termed as apparent undigested residue. Use a representative sub-sample of this preparation for analyses. In order to get a representative sub-sample, it is advised to grind the entire residue in a pestle and mortar or in a small rotating-ball grinding mill.

11.5.2. Preparation of lyophilized rumen microbial fraction

Two hours after the morning feed, collect about 1 litre rumen liquor from a cow fed a roughage-based diet. Pass the liquor through two layers of muslin cloth and then keep at 4°C for 30 min in a carbon dioxide-flushed cylinder of 1 litre capacity. Separate the rumen fluid devoid of heavy and light particles by pipetting the liquor from the cylinder between the heavy particles which settle down and the light particles which float on the top. This method for collection of rumen fluid is essentially adapted from Yang and Russell [22]. Centrifuge (20,000 *g,* 20 min, 4°C) several portions (each 30–35 ml) of this liquor. Wash the pellets with distilled water followed by centrifugation (20,000 g, 20 min, 4°C). Repeat this washing step two more times, lyophilize the pellets and pool them. Use a sub-sample of one preparation. The nitrogen content of this fraction should be approximately 7% [23]. In the lyophilized rumen microbial fraction prepared in our laboratory in the manner as described above, the nitrogen content was 7%.

11.5.3. Determination of purines (marker for microbial mass/protein)

Spectrophotometric methods
(According to Zinn and Owens [24], with some modification as described in Makkar and Becker [21]).

- Reagents
  - *Ammonium dihydrogen phosphate (0.2 M):* Dissolve 23 g ammonium dihydrogen phosphate in about 700 ml distilled water, and then make up the volume to 1 litre with distilled water.
  - *Sodium hydroxide (10 M):* Dissolve 40 g NaOH in approximately 70 ml distilled water and then make up the volume to 100 ml with distilled water.
  - *$AgNO_3$ (0.4 M):* Dissolve 1.6987 g $AgNO_3$ in approximately 15 ml distilled water and then make up the volume to 25 ml with distilled water. Protect the solution from light. Store in a brown bottle and surround the bottle with black paper.
  - *HCl (0.5 M):* Dilute 10 ml HCl (37%) to 240 ml with distilled water.
  - *Ammonium dihydrogen phosphate (28.5 mM):* Dilute 100 ml of 0.2 M solution of ammonium dihydrogen phosphate to 700 ml with distilled water.
  - o-*Phosphoric acid (85%).*

- *Perchloric acid (0.6 M):* Dilute 10 ml perchloric acid (70%; 12 M) to 200 ml with distilled water.

➢ Procedure

Weigh the lyophilized microbial fraction or the apparent undigested residue (25–75 mg) in 25 ml screw-cap tubes and add 2.5 ml of 0.6 M perchloric acid. Incubate the mixture in a water bath at 90–95°C for 1 h. After cooling, add 7.5 ml of 28.5 mM ammonium dihydrogen phosphate and return the tubes to a water bath (90–95°C) for 15 min. After cooling, centrifuge (3,000 *g,* 10 min) the contents and collect the supernatant. Add an aliquot (0.25 ml) of the supernatant to 4.5 ml of 0.2 M ammonium dihydrogen phosphate and adjust the pH between 2 and 3 (generally to 2.5) using o-phosphoric acid. After the pH adjustment (usually 5 µl o-phosphoric acid gives pH 2.3), add 0.25 ml of $AgNO_3$ (0.4 M) and keep the mixture overnight at 5°C in the dark. Centrifuge (3,000 *g,* 10 min) the contents and discard the supernatant. Take care not to disturb the pellet. Wash the pellet with 4.5 ml distilled water adjusted to pH 2 (with sulphuric acid) followed by centrifugation. Suspend the pellet in 5 ml of 0.5 M HCl, vortex thoroughly and transfer to the 90–95°C water bath for 30 min after covering the tubes with marbles. Centrifuge (3,000 *g,* 10 min) the tubes and record absorbance of the supernatant at 260 nm against 0.5 M HCl. For studies with RNA in the range of 25–75 mg (instead of lyophilized microbial preparation or the apparent undegraded residue), read the absorbance at 260 nm after 1:10 dilution of the supernatant.

Without adjustment of the pH (which is generally 3.4) to between 2 and 3 before addition of the $AgNO_3$ solution (see below), the recovery of purine basis from yeast RNA (Sigma) is generally lower (80–90% vs. 94–99%), suggesting the importance of the pH-adjustment step in obtaining satisfactory recoveries. Addition of the $AgNO_3$ solution does not change the pH of the solution. Use o-phosphoric acid for adjustment of pH to 2.7.

Express results either based on RNA or lyophilized microbial preparation.

➢ HPLC method for determination of purine bases (adenine and guanine)

(According to Balcells *et al.* [25], with some modifications as described in Makkar and Becker [21]).

➢ Equipment, reagents, HPLC conditions and analysis

The HPLC equipment, which we used, consisted of a Merck Hitachi L-7100 HPLC pump, an L-7450 photo diode array detector, an L-7200 autosampler, a D-700 interphase module and an LC organiser.

Analytical column: Reverse phase C18 (LiChrospher 100, endcapped 5 µm) 250 mm x 4 mm I.D. (Lichrocart; Merck, Darmstadt, Germany) protected by a guard column containing the material as in the main column.

HPLC solvent A: 10 mM $NH_4H_2PO_4$ and adjust pH to 6 with 10% NH_4OH.

(Dissolve 11.503 g $NH_4H_2PO_4$ in about 500 ml distilled water and then make the volume to 1 litre with distilled water. It is 100 mM solution. Pipette 100 ml of this solution and dilute to 1 litre to obtain 10 mM solution)

HPLC solvent B: Add 150 ml of acetonitrile to 600 ml of 12.5 mM $NH_4H_2PO_4$ (75 ml of 100 mM solution plus 525 ml of distilled water) and adjust pH to 6 with 10% NH_4OH.

(Filter solvents A and B through a 0.45 µm filter and degas by ultrasonication and by application of vacuum).

Purine bases and internal standard solution for converting integration units to the concentration: Prepare 1 mM stock solution. Put a measuring flask (250 ml capacity) containing approximately 50 ml distilled water on a magnetic stirrer fixed with a hot plate. Add to the flask 100 µl of 10 M sodium hydroxide solution. Heat at about 90°C and then transfer 33.77, 37.77 and 34.03 mg of adenine and guanine and allopurinol respectively to the flask. Wait (generally 30 min) till the contents dissolve. Cool the contents and make up the volume to 250 ml with distilled water. This solution can be stored at 4°C for 10 days. Dilute this stock solution 12.5 times; pipette 2 ml of the stock solution into 25 ml measuring flask and make up the volume to 25 ml with buffer A of the HPLC. Inject 40 µl of this solution into the HPLC.

Preparation of 3 mM allopurinol solution: Take 100 ml measuring flask and weigh-in 40.83 mg allopurinol. Add approximately 50 ml of distilled water and 20 µl of 12 N (70%) perchloric acid. Heat the contents to approximately 90°C with stirring on a magnetic stirrer. Cool the contents to room temperature and make up the volume to 100 ml with distilled water.

Preparation of 8 mM caffeine solution: Dissolve 155.36 mg caffeine in 80 ml of distilled water and then make up the volume to 100 ml with distilled water.

Gradient: A 30-min linear gradient from 0 to 100% solvent B. After 40 min, increase solvent A to 100% in the following 5 min and equilibrate the column to the starting condition (100% A) in the next 15 min before injecting the next sample.

Time (min)	Solvent A (%)	Solvent B (%)
0.0	100	0
30	0	100
40	0	100
45	100	0
60	100	0

Detection wavelength: 254 nm with a full scale deflection set at 0.2 absorbance

Column temperature: Ambient (approximately 22°C)

Guanine and adenine appear at about 11 and 15.5 min respectively. Allopurinol or caffeine can be used as internal standards. These appear at about 13.5 and 29.5 min respectively. For tannin-containing feeds, do not use caffeine since it binds with tannins, which lowers the recovery of caffeine [21].

➢ Sample hydrolysis

Weigh 25–100 mg sample in 25 ml screw-cap tubes and add 2.5 ml of perchloric acid (0.6 M) and 0.5 ml of an internal standard (3 mM allopurinol or 8 mM caffeine). Incubate the mixture in a water bath at 90–95°C for 1 h. After cooling, add 7.0 ml of Buffer A of the HPLC system, adjust the pH between 6.6 and 6.9 using concentrated KOH (approximately 8 M) and freeze overnight. Next day thaw the sample and then centrifuge (approx. 10,000 *g*, 10 min) to remove the precipitate formed (freezing overnight before the centrifugation step improves the removal of suspended material). Filter through 0.45 μm filter and inject appropriate volume (15–50 μl) into the HPLC.

Express results based on adenine plus guanine.

➢ Some useful data on purines (Tables XI and XII).
(According to Makkar and Becker [21])

Table XI. Purine determination (as $A_{260\ nm}$) using spectrophotometric method

	Absorbance at 260 nm	
	Mean	S.D.
25 mg LRM	0.215	0.006
50 mg LRM	0.456	0.005
75 mg LRM	0.675	0.005
25 mg RNA*	0.210	--
50 mg RNA*	0.414	--

LRM, lyophilized rumen microbes
A260 nm = (0.009207).mg LRM – 0.01178 ($r^2 = 0.99$; n = 3)
* Average of two values and after 1:10 dilution

Table XII. Purine base determination using HPLC Method

	Adenine (μmol)		Guanine (μmol)	
	Mean	SD (n=3)	Mean	SD (n=3)
50 mg LRM	2.35	0.03	2.95	0.02

LRM, lyophilized rumen microbes

Note: It may be noted that workers can optimize their own system for a particular material being studied once equipped with a sound grasp of the rationale behind these procedures

12. RADIOLABELLED BSA PRECIPITATION METHOD

This is the most sensitive method for determining amounts of protein precipitated by a tannin. It is also useful for running competitive binding assay to characterize tannin-protein interactions. The method described here [26] has been scaled down from the original method [27].

Caution: Proper handling of isotopes is required for this method. You must be approved for isotope use by the appropriate committee of your institution.

12.1. Reagents

(1) *pH 4.9 acetate buffer (Buffer A):* Acetic acid (0.20 M), NaCl (0.17 M), pH adjusted to 4.9 with NaOH (11.4 ml glacial acetic acid, 9.86 g NaCl dissolved in about 800 ml water, then adjust to pH 4.9 with a solution of NaOH, then bring to a final volume of 1 liter).
To make large volumes of Buffer A conveniently, prepare the following

2 M acetic acid, 1.7 M NaCl. Add 114 ml glacial acetic acid to about 800 ml distilled water, add 99.4 g NaCl, and bring to 1 litre with distilled water. Store in refrigerator.

2 M sodium acetate, 1.7 M NaCl. Add 164.1 g sodium acetate to about 800 ml distilled water. Add 99.4 g NaCl and bring to 1 litre with distilled water. Store in refrigerator.

Buffer A: Mix 40 ml of the acetic acid solution with 60 ml of the sodium acetate solution and bring to 1 litre. Check the pH; it should be 4.9. (store Buffer A in refrigerator)

(2) *Radioiodinated bovine serum albumin.* For the groups participating in the FAO/IAEA Co-ordinated Research project, IAEA obtained radioiodinated BSA from ICN, catalogue number 68031. This is radioiodinated by the chloramine T method. The radiolabelled BSA was procured in the lyophilised state and each vial had 41.7 μCi of radiolabelled BSA, which was air transported to the groups. The specific radioactivity of the sample was 0.85 mCi/mg. The groups were asked to follow the following procedure.

Soon after arrival of the vial (containing the lyophilized radiolabelled BSA; 41.7 μCi), add 2.5 ml of unlabelled 2 mg/ml BSA in buffer A. Dissolved it by shaking gently, and then dispensed into plastic vials in 200 μl aliquots and freeze them. The 200 μl aliquots to be used on a given day should be thawed, diluted 10-times with unlabelled 2 mg/ml BSA in buffer A and immediately dialyzed for 1 to 2 h at 4°C against 500 ml buffer A. Thus for a typical experiment 2 vials of labelled protein might be used; the total 400 µl of labelled protein would be diluted to 4 ml with unlabelled BSA and then dialyzed. After dilution, one should get around 30,000 cpm in a 50 µl aliquot of protein (although this of course depends somewhat on time since labelling and decay).

Note: The labelled BSA in the lyophilised form was procured on special request. The labelled BSA (Catalogue number 68031 from ICN) is normally supplied in liquid state. If that is the case, each vial will have 150 µl labelled BSA of specific activity 0.85 mCi/mg. Dilute it 100-fold with unlabelled 2 mg/ml BSA in buffer A. i.e., 150 µl provided by ICN is diluted to a total volume of 15.0 ml. Then dispense this solution into 200 µl aliquots and freeze for later use. On a given day, thaw the 200 µl aliquots and dilute it 10-times with unlabelled 2 mg/ml BSA in buffer A and immediately dialyse for 1–2 h at 4°C against buffer A. Rest of the procedure is the same as mentioned above for the lyophilised labelled BSA.

(3) *Unlabelled ("carrier") BSA*: 2.0 mg/ml in Buffer A. Prepare weekly and store in a refrigerator. Use to dilute the labelled protein to a suitable activity and as a carrier in the TCA-precipitability controls.

(4) *Plant Extracts and Tannic acid Standard:* Prepare the tannic acid (Merck) standard at 0.1 mg/ml. It should be initially dissolved at 1.0 mg/ml in 50% aqueous methanol, and then diluted 10-times with 50% aqueous methanol. Plant extracts are prepared in 50% aqueous methanol as for protein-precipitable phenolics (see Section 2.2.4.), and are then diluted 5 or 10 times with 50% aqueous methanol as needed to obtain linear response.

(5) *Trichloroacetic acid (100%):* Commercially purchased 100% (~6.1 N) TCA. This solution is stable indefinitely. More dilute solutions of TCA cannot be used reliably after prolonged storage.

12.2. Supplies

(1) 0.5–0.65 ml microfuge tubes (less expensive tubes which can withstand 13K *g* centrifugation but can easily be cut). For example, Life Science Products 8507-GMT 0.65 ml Bes Tubes.

(2) Tube opener (for example, Life Science Products TT-44 Tube opener).

(3) Tube cutter (for example, Life Science Products TC018-A cutter and TC018-B replacement blade.
(Life Science Products, http://www.lifesciprod.com, Fax +1-303-4527689. Similar products are available from many other suppliers e.g. Fisher)

(4) Also needed are gloves and other isotope work area supplies; microfuge; gamma counter.

12.3. Procedure for BSA precipitation (amount of BSA precipitated by tannic acid or plant extracts)

The assay is run in 0.5 ml microfuge tubes. Buffer A, labelled protein, and water (as needed to attain a final total volume of 400 µl) are added to the tubes, which are then capped and vortexed. The tannin or plant extract is then added with immediate vortexing. The samples are incubated overnight at 4°C and are then centrifuged for 5 min in a microfuge (13,000 *g*) at room temperature. The supernatants are carefully aspirated without disrupting the pellets. The surface of the pellet is gently washed with 100 µl buffer, the samples centrifuged again for 1 min, and the solution aspirated. The tubes are trimmed to fit into the gamma counter vials, and the pellets are then counted in a gamma counter directly, without transferring them out of the microfuge tubes.

Assay tubes contain:

(1) 50 µl labelled protein
(2) 250 µl buffer A
(3) 100-0 µl 50% aqueous methanol
(4) 0-100 µl tannic acid solution or plant extract

For each assay, the following controls are run:

(1) Total counts used per aliquot labelled protein.
(2) Background: 100 µl water, no tannin. This value is subtracted from all values for counts precipitated. We usually find that the background is less than 10% of the counts added.
(3) TCA precipitable counts: 50 µl labelled protein plus 190 µl "carrier" unlabelled BSA (2 mg/ml) mixed with 60 µl of 100% TCA and incubated with the samples at 4°C (15 min is sufficient). It is desirable to use labelled protein in which >90% of the counts are TCA-precipitable.

12.4. Calculations

The amount of protein precipitated is calculated directly from the counts in the precipitate after correcting for background (non-specific) precipitation. The initial concentration of the labelled BSA (50 µl of 2.0 mg/ml = 100 µg) is used for the calculation. The amount of protein precipitated is plotted as a function of the dry matter equivalent of plant extract, and each plant extract is tested at ≥ 3 concentrations. The slope of the linear portion of the curve gives the protein precipitation capacity of the plant, as µg protein precipitated per mg forage. The capacity of the tannic acid standard is determined each time the assay is run for inter- and intra-lab quality control.

13. SIMPLIFIED RADIOLABELLED BSA PRECIPITATION METHOD

The above radiolabelled method has been simplified to eliminate the more difficult procedures of relatively high speed (13,000 *g*) centrifugation and manipulations of small volumes of radiochemical solutions by binding the radiolabelled protein to tannin immobilized on a paper disk. The amount of radioactivity on the paper disk is determined by gamma counting [28].

13.1. Reagents

(1) *pH 4.9 acetate buffer (Buffer A)*: See Section 12.1.

(2) *Radioiodinated bovine serum albumin:* Use the radiolabelled BSA as prepared in Section 12.1. i.e., after dialysis and dilution and having around 30,000 cpm in a 50 μl aliquot (300 cpm/μg protein). Note that in preparing this solution all dilutions have been made with unlabelled BSA of 2 mg/ml in Buffer A.

(3) *Plant Extracts:* Plant extracts are prepared in 50% aqueous methanol as for protein-precipitable phenolics, and are then diluted 5x or 10x with 50% aqueous methanol as needed to obtain linear response.

13.2. Procedure for protein binding

13.2.1. Immobilization of tannins from the plant extract

Whatman #1 filter paper disks (1 cm diameter, Fisher Scientific, Pittsburg, PA) were used as the solid phase. The plant extracts or standard solutions were dispensed onto the disks using a micropipette or a 5 μl microcapillary tube. Successive 5 μl aliquots were dispensed, with brief drying at room temperature after each aliquot, to achieve the desired concentration of immobilized tannin. Blank disks, treated with the extracting solvent (50% aqueous methanol), were also prepared. *Caution*: If larger volumes are applied, or if successive aliquots are applied too rapidly, the disks become too wet and some tannin is lost.

13.2.2. Protein binding on immobilized tannins

Aliquots (25-100 μl) of the radiolabelled BSA solution (Section 12.2) were pipetted into counting vials for calibration of the method, and the remainder of the 10 ml of protein solution was placed in a flat dish (Petri dish). The tannin-loaded disks were then exposed to a bulk solution of the radiolabelled protein. Up to 20 paper disks with immobilized tannins were added to the protein solution carefully so that individual disks did not overlap or stick together. The disks were then gently agitated on a platform shaker for 30 min at room temperature. This protocol simplifies the mechanics of the protein binding method by eliminating the high speed centrifugation and manipulation of small volumes.

The protein solution was poured off the disks into a suitable waste container, and was replaced thrice with the acetate buffer; each time the disks were gently agitated for 30 min at room temperature. The buffer was discarded as low-level radioactive waste each time. The disks were then placed in counting vials for gamma counting. The amount of protein bound to each disk was calculated using the calibration curve after subtracting the amount of protein bound by the blank (tannin-free disk).

Notes:

The reaction between immobilized procyanidin and protein was complete after about 10 min of incubation. Longer incubations (30 min) were routinely used to ensure that equilibrium binding would be achieved for the mixture of polyphenolic components in unfractionated plant extracts. Although there was some non-specific binding of protein by the paper disks, three washing steps reduced non-specific binding to about 25 ± 3 μg protein per disk. This background was determined each time the method was performed so that the background binding could be subtracted from experimental values.

In the linear region, the immobilized procyanidin bound 8.9 ± 1.2 μg protein/μg procyanidin. When procyanidin was reacted with excess BSA in solution, the precipitates isolated by centrifugation comprised of about 8 μg protein/μg procyanidin. The quantitative similarity between results obtained with procyanidin in solution and procyanidin immobilized on a paper disk suggests that the immobilization does not alter the reaction between procyanidin and protein.

When small amounts of procyanidin were applied to the disks, the amount of protein bound to the disks increased linearly as expected since protein is in excess. However, protein binding was independent of procyanidin concentration when large amounts of procyanidin were loaded on the disks. The limited surface area of the disks may limit the accessibility of immobilized polyphenolic for reaction with protein. The immobilized polyphenolic method is useful for determining the protein binding capacity of tannins only in the linear range, where protein binding is dependent on the amount of polyphenolic extract loaded onto the disk.

Several types of disks were tested for immobilization of tannin. Cellulose acetate binds large amounts of protein even in the absence of sorbed polyphenolic, so was not suitable for the protein binding method. Paper disks were convenient to work with, inexpensive, and did not interfere with the analytical methods.

14. CHARACTERIZATION OF PHENOLIC COMPOUNDS BY THIN LAYER CHROMATOGRAPHY (TLC)

(According to Mueller-Harvey *et al* and Mueller Harvey [29, 30])

14.1. Characterization of condensed tannins

14.1.1. Samples

About 500 g of the shade dried plant material should be ground first to pass a 2 mm screen. All the ground material including those parts remaining inside the mill should be taken, mixed well and approximately 100 g of this sample is again ground to pass through a 0.5 mm screen. Take care that at any stage of the grinding, the sample temperature does not rise above 40°C.

14.1.2. Reagents

(1) *70% acetone*: Mix 70 ml acetone (Analar grade) and 30 ml distilled water.

(2) *Butanol/HCl reagent*: Thoroughly mix 95 ml butan-1-ol (Analar grade) and 5 ml HCl (12 M).

(3) *Anthocyanidin standards*: Chlorides of cyanidin, delphinidin, pelargonidin, etc (only the first two are very common, but others may also be present and can be seen on the plates) can be purchased from Apin Chemicals, Abingdon, UK, Fax: +44-1235-83200, Tel: +44-1235-832515 or Extrasynthese, 69730 Genay, France, Fax: +33-478-981945, Tel: +33-478-982034 or Fluorochem, Old Glossop, UK., Fax: +44-1457-869360, Tel: +44-1457-868921.

14.1.3. Material for TLC

(1) *Cellulose MN300 plates:* Polygram CEL 300, cat. no. 801 013, Macherey and Nagel, Dueren, Germany; Tel +49-2421-9690, Fax: +49-2421-969199; order the 20 x 20 cm plates (these have plastic at the back) and cut them in a guillotine, soft side down, to 10 x 10 cm or even 5 x 5 cm if you want to shorten the time.

(2) *Disposable micro-pipettes*: Camlab, UK or any other TLC supplier; recommended sizes 1, 2 and 5 µl.

(3) *TLC sprayers which produce a fine mist (need a fine nozzle) and are resistant to 12 M HCl*: e.g. Merck TLC sprayer cat. no 370/2000/10 (approx £200).

(4) This is a good sprayer but is very expensive. Glass sprayers can also be used if the nozzle is fine (Merck ca. no. 370/0642/00 or 306/6261/00. The pressure relief vent of the latter assembly is recommended as the simple bulb of the first assembly is more likely to produce an uneven spray. However, the reagent spray atomizer is likely to give the finer mist of the two). It is vital that all glass sprayers are cleaned with distilled water immediately after use, as the nozzles become easily blocked.

(5) *TLC solvents*: Two solvents are required.

Solvent 1 (for first direction): Mix 100 ml concentrated formic (85%), 10 ml 12 M HCl and 30 ml water.

Solvent 2 (for second direction): Mix 20 ml pentan-1-ol, 10 ml glacial acetic acid and 10 ml water.

14.1.4. Extraction

Leaf samples (200 mg of dried plant material passed through a 0.5 mm screen) are taken in a glass beaker of approximately 25 ml capacity. To it is added 10 ml of aqueous acetone (70%) and the beaker is suspended in an ultrasonic water bath (Branson 3210) and subjected to ultrasonic treatment for 20 min at room temperature. The contents of the beaker are then transferred to centrifuge tubes and subjected to centrifugation for 10 min at approximately 3,000g at 4°C (if refrigerated centrifuge is not available, cool the contents by keeping the centrifuge tube on ice and then centrifuge at 3,000 *g* using an ordinary clinical centrifuge). Collect the supernatant and keep it on ice.

Take 1 ml of the aqueous acetone extract and evaporate the solvent to less than 200 μl in a stream of nitrogen (not oxygen, as this would oxidise the tannins!). Dilute the concentrate to 400 μl with water and mix well. Take an aliquot of 80 μl from this aqueous solution and add 1 ml of the HCl/butanol reagent, cover the tubes (marbles or loose Teflon lined Pyrex tube screw tops) heat at 100°C for 60 min. Cool the solution and spot aliquots onto the TLC plates.

14.1.5. Procedure

In a fume cupboard, fill two small TLC tanks with solvents 1 or 2 to a height of about 3 mm, ensure that the atmosphere in the tanks is well saturated with the solvent (a sheet of filter paper can be dipped in the solvent and surround around the tank walls inside to speed up the saturation process, 'seal' the lids with grease to prevent evaporation of the solvent; the TLC plate should not be run before 30 min after filling the tanks with the solvents).

Spot the sample carefully at the bottom left corner (7 mm from the edges). The diameter of the spot should not exceed 5–7 mm. The volume of sample to be spotted depends on the concentration of the anthocyanidins. It is best to try out a range of volumes (e.g. 5, 10, 20 μl) by repeatedly applying 5 μl onto the same spot and letting the spot dry between applications. Do not use hot air blower/hair drier to dry the spots.

Gently lower the plate into the TLC tank. Switch off the fume cupboard to prevent drafts crossing the tank and causing temperature gradients. Remove the plate when the solvent front has just reached the top of the TLC plate (approx. 3 mm below the top); but switch the fume cupboard on again for this. Dry the plate in the draft of the fume cupboard (do not use hot air). When all solvent has evaporated, turn the plate by 90 degrees and repeat the separation with solvent 2. The colours of the anthocyanidins should be clearly visible. They are identified from their position (Rf-values) on the plate [see below] and by their characteristic colours.

14.2. Survey of condensed and hydrolysable tannins by TLC

14.2.1. Solvents

Solvent 1: Mix 2 ml glacial acetic acid and 98 ml water.

Solvent 2: Mix 60 ml butan-1-ol, 15 ml glacial acetic acid and 25 ml water.

14.2.2. Spray reagents to detect different classes of tannins

Some of these sprays require some skill in order to extract good and reproducible information from them. The best thing, therefore, is to keep trying until one is familiar with the colour reaction conditions. A lot of information can be obtained by the combination of different sprays and the positions of the spots on the TLC plates. Note that some spots will react with several of these sprays. For example, condensed tannins will react positively with spray 1 and 2. Hydrolysable tannins (i.e. gallotannins) will react with sprays 1 and 3, and ellagitannins with sprays 1 and 4. Catechin gallate type tannins will react with sprays 1, 2 and 3 (e.g. present in *Acacia nilotica*). The position on the TLC plate also provides information, some tannins will hardly move away from the origin where the plant extract was applied. These tannins are likely to be of higher molecular weight and this information alone might be useful in interpreting some animal responses.

(1) *Vanillin/HCl reagent:* Every day, prepare freshly a solution containing 1 g vanillin in 10 ml 12 M HCl. Be careful when spraying this reagent, it is very corrosive. Cut one side out of a large cardboard box, place it in a fume cupboard. Activate the spray, test that the droplets are finely distributed on a piece of paper, then pass this mist over the TLC plates. You will need to gauge yourself how much spray to apply per plate, initially just apply the minimum amount until red spots start to appear.

This reagent works very well and detects flavan-3-ols, e.g. catechin and epi-catechin, plus condensed tannins. When these are present, pink spots are obtained. You will need to keep a record of these red spots. This can be done by using a sharp tool (edge of thin spatula, scalpel or razor blade) to surround each spot with a series of small holes (recommended option).

(2) *Photocopying the TLC plate*: Be careful while photocopying, the acid might damage the photocopier; consider covering it with a transparent film first or placing some tracing paper over the TLC plate and copying the spots. TLC surface is quite fragile, so care should be exercised while handling the plate.

(3) *Ferric ion reagent:* Prepare daily a fresh solution containing 1 g of $K_3Fe\ (CN)_6$ and 1 g $FeCl_3$ in 50 ml water. Then add 5 tiny crystals of $KMnO_4$. Lightly spray the TLC plate with a fine mist of this reagent (avoid spluttering large drops on the plate). The background of the plate tends to turn dark blue more or less quickly. This can be reduced by laying the TLC plate (Cellulose surface point up) into a glass basin containing 2 to 3 M HCl soon after applying the reagent {Safety note: HCN is released, therefore, use a fume cupboard}. Once the plates are thoroughly rinsed in the HCl bath {careful not to damage the cellulose surface}, they are transferred to a second basin filled with water and left to soak for several hours. Ideally, blue spots are obtained on a relatively white background – this requires a bit of practice.

This reagent works well once you have sorted out the conditions, i.e. how much spray to apply. It is a general reagent for all phenolic compounds (tannins and others). If the sample has large quantities of hydrolysable tannins, these often appear as large spots. It is a useful spray, as it provides guidance for the next two sprays which can be more difficult.

(4) *Potassium iodate reagent*: Prepare a saturate solution of KIO_3 (potassium iodate) in distilled water (i.e. add enough KIO_3 crystals until some of them no longer dissolve in the water).

This reagent detects gallic acid and its esters (i.e. gallotannins, a subgroup of the hydrolysable tannins). It is suggested that this reagent is tried initially using gallic acid as a standard. The coloration is orange pink to brown and does not always appear clearly. Therefore, several attempts will be needed. Sometimes it is helpful to hold the TLC plates towards the window up to eye-level and to look across the white surface. This has the effect of 'concentrating' the colour and makes it easier to see the spots.

(5) *Sodium nitrite reagent:* Cool 10 ml of water to near 0°C, add 20 mg of $NaNO_2$ (sodium nitrite) plus 1–2 drops of glacial acetic acid.

This reagent detects ellagic acid and its esters. Orange-brown spots are obtained. This is a tricky reagent and it will require several attempts to become confident with it. Again, try it out first using ellagic acid as the standard.

14.3. TLC procedure

Apply the aqueous acetone extracts (5 to 30 μl) to several cellulose MN300 TLC plates (10 x 10 cm) and try to keep the spot at the origin as small as possible without damaging the TLC surface (approximately 7 mm diameter). Place the TLC plate in 'solvent 1' first and remove it when the solvent front has just reached the top of the plate. Dry in a cool stream of air, turn the plate by 90 degrees and place the plate into a tank containing 'solvent 2'.

Then subject each TLC plate to one of the 4 spray reagents separately.

15. REFERENCES

[1] MAKKAR, H.P.S., BLUEMMEL, M., BOROWY, N.K., BECKER, K., Gravimetric determination of tannins and their correlations with chemical and protein precipitation methods, J. Sci. Food Agric. **61** (1993) 161–165.

[2] MAKKAR, H.P.S., BECKER, K., Behaviour of tannic acid from various commercial sources towards some chemical and protein precipitation assays, J. Sci. Food Agric. **62** (1993) 29–299.

[3] MAKKAR, H.P.S., BLUEMMEL, M., BECKER, K., Formation of complexes between polyvinyl pyrrolidone and polethylene glycol with tannins and their implications in gas production and true digestibility in *in vitro* techniques, Brit. J. Nutr. **73** (1995) 897–913.

[4] PORTER, L.J., HRSTICH, L.N. CHAN, B.G., The conversion of procyanidins and prodelphinidins to cyanidin and delphinidin, Phytochemistry **25** (1986) 223–230.

[5] MAKKAR, H.P.S., BECKER, K., Vanillin-HCl method for condensed tannins: effect of organic solvents used for extraction of tannins, J. Chem. Ecol. **19** (1993) 613–621.

[6] MAKKAR, H.P.S., BECKER, K., Some problems in the determination of tannins and possible solutions, Acta Horticultura **381** (1994) 782–788.

[7] INOUE, K.H., HAGERMAN, A.E., Determination of gallotannins with rhodanine, Anal. Biochem. **169** (1988) 363–369.

[8] MAKKAR, H.P.S., BECKER, K., Gallotannin determination using HPLC, Unpublished.

[9] HARTZFELD, P.W., FORKNER, R., HUNTER, M.D., HAGERMAN, A.E., Determination of hydrolysable tannins (gallotannins and ellagitannins) after reaction with potassium iodate, J. Agric Food Chem. **50** (2002) 1785–1790.

[10] HASLAM, E., Galloyl esters in the Aceraceae, Phytochemistry **4** (1965) 495–498.

[11] BATE-SMITH, E.C., Astringent tannins of *Acer* species, Phytochemistry **16** (1977) 1421–1426.

[12] WILLIS, R.B.; ALLEN, P.R., Improved method for measuring hydrolysable tannins using potassium iodate, The Analyst **123** (1998) 435–439.

[13] MAKKAR, H.P.S., DAWRA, R.K., SINGH, B., Determination of both tannin and protein in a tannin-protein complex, J. Agric. Food Chem. **36** (1988) 523–525.

[14] MAKKAR, H.P.S., DAWRA, R.K., SINGH, B., Determination of both tannin and protein in a tannin-protein complex, J. Agric. Fod Chem. **36** (1988) 523–525.

[15] HOFFMANN, E.M., MUETZEL S., BECKER, K., A modified dot-blot method of protein determination applied in the tannin-protein precipitation assay to facilitate the evaluation of tannin activity in animal feeds, Brit. J. Nutr. **87** (2002) 421–426.

[16] DAWRA, R.K., MAKKAR, H.P.S., SINGH, B., Protein binding capacity of microquantities of tannins, Anal. Biochem. **170** (1988) 50–53.

[17] HAGERMAN, A.E., Radial diffusion method for determining tannins in plant extracts, J. Chem. Ecol., J. Chem. Ecol. **13** (1987) 437–449.

[18] TILLEY, J.M., TERRY, R.A., A two-stage technique for the *in vitro* digestion of forage crops, J. Brit. Grasslands Soc. **18** (1963) 104–111.

[19] MENKE, K.H., RAAB, L., SALEWSKI, A., STEINGASS, H., FRITZ, D., SCHNEIDER, W., The estimation of the digestibility and metabolizable energy content of ruminant feedstuffs from the gas production when they are incubated with rumen liquor *in vitro*. J. Agric. Sci. **93** (1979) 217–222.

[20] GATECHEW, G., MAKKAR H.P.S., BECKER, K., Stoichiometric relationship between short chain fatty acid and *in vitro* gas production in presence and absence of polyethylene glycol for tannin containing browses, gas production: fermentation kinetics for feed evaluation and to assess microbial activity, An EAAP Satellite Symposium, Wageningen, 18–19 August (2000) 46–47.

[21] MAKKAR, H.P.S., BECKER, K., Purine quantification in digesta from ruminants by spectrophotometric and HPLC methods, Brit. J. Nutr. **81** (1999) 107–113.

[22] YANG, C-M.J., RUSSELL, J.B., Resistance of proline-containing peptides to ruminal degradation *in vitro*, Appl. Environ. Microbiol. **58** (1992) 3954–3958.

[23] ORSKOV, E.R., Rumen microorganisms and their nutrition, In: Protein Nutrition in Ruminants (ORSKOV, E.R., Ed.), (1982) 19–40.

[24] ZINN, R.A., OWENS, F.N., A rapid procedure for purine measurement and its use for estimating net ruminal protein synthesis, Can. J. Anim. Sci. **66** (1986) 157–166.

[25] BALCELLS, J., GUADA, J.A., PEIRÓ, J.M., Simultaneous determination of allantoin and oxypurines in biological fluids by high-performance liquid chromatography, J. Chromatography **575** (1992) 153–157.

[26] HAGERMAN, A.E., RICE, M.E., RITCHARD, N.T., Mechanisms of protein precipitation for two tannins, pentagalloyl glucose and epicatechin$_{16}$ (4→8) catechin (procyanidin), J Agric Food Chem **46** (1998) 2590–2595.

[27] HAGERMAN, A.E., BUTLER, L.G., Specificity of proanthocyanidin-protein interactions, Biol. Chem. **256** (1981) 4494–4497.

[28] HENSON, G.L., NIEMEYER, L., ANSONG, G., FORKNER, R., MAKKAR, H.P.S., and HAGERMAN, A.E., Modified method for determining protein binding capacity of plant polyphenolics using radiolabeled protein. Phytochemical Analysis (2003), in press.

[29] MUELLER-HARVEY, I., REED, J.D., HARTLEY, R.D., Characterisation of phenolic compounds, including flavonoids and tannins, of 10 Ethiopian browse species by high performance liquid chromatography, J. Sci. Food Agric. **39** (1987) 1–14.

[30] MUELLER-HARVEY, I., Analysis of hydrolysable tannins, Anim. Feed Sci. Technol. **91** (2001) 3–20.

16. A SUMMARY OF RESULTS FROM THE FIRST PHASE OF THE CO-ORDINATED RESEARCH PROJECT

This CRP was initiated in 1999. The Second Research Coordination Meeting was held from 19 to 23 November 2001 at Centro de Energia Nuclear na Agricultura (CENA), Universidade de Sao Paulo, Piracicaba, Brazil, to: a) review the results obtained so far on development, refinement, standardisation and validation of tannin assays to seek correlation with animal performance indicators; and b) plan future studies. It was attended by 10 Research Contract and Agreement holders, the Scientific Secretary of the Project, and 10 staff members and students of the host institution.

16.1. The tannin assays and standards used in the first phase

Assays	Suggested standard
Gallotannins by rhodanine method	Gallic acid
Gallotannins by HPLC	Gallic acid
Condensed tannins (butanol-HCl-iron method)	Anthocyanidin (leucocyanidin)/cyanidin chloride
Total phenols and total tannins (Folin-Ciocalteu method)	Tannic acid
Protein precipitable phenolic assay	Tannic acid
Radial diffusion assay	Tannic acid
Filter paper-BSA-Ponceau S assay	Tannic acid (or as mg bovine serum albumin, fraction V bound/mg dry leaf sample)
Radio-labelled protein binding assay	Tannic acid (or as mg bovine serum albumin, fraction V bound/mg dry leaf sample)

It was suggested that all groups should use tannic acid from Merck. This was decided based on a study wherein tannic acid from various commercial sources were compared. The data showed: a) lowest amount of free gallic acid in tannic acid, and b) highest protein precipitation capacity, and lowest threshold level at which precipitation of bovine serum albumin started by tannic acid from Merck. In addition, batch-to-batch variation was not significant in tannic acid from this source between the two tested batches.

A tannin bioassay based on measurement of both gas and microbial mass production on incubation of tanniniferous foliage in the *in vitro* gas method in presence and absence of polyethylene glycol (MW 4,000/6,000) was also used in the project. The suggested time of incubation was 16 and 24 h.

16.2. Conclusions

- All Research Contract holders established the above-mentioned tannin assays, including the ^{125}I-labelled BSA method in their laboratories and used them for characterization of tree leaves and browses for tannin levels and activity. All participating groups used the same protocols and standards for tannin assays, which enabled comparison of the results and provided meaningful information.
- A total of 48 species of tree leaves and browses were characterized for tannin levels and activity. Ten of them were also evaluated in *in vivo* studies. The apparent digestibility coefficients of N correlated best with the following tannin assays: total phenol, total tannins, condensed tannins, radiolabelled BSA method, and percentage increase in gas on inactivation of tannins using polyethylene glycol (PEG) in the *in vitro* method for measuring gas production. When these tannin values were adjusted according to the N level in the forages (tannin% / N%), correlations remained high but were not improved. None of these values was a good predictor of feed intake in the short term *in vivo* studies. Under the conditions of these short-term studies, using a seven day adaptation period, the intake data should be interpreted with caution. The intake data gave variable relative values but there were indications that tanniniferous forages need a longer adaptation period.
- Total phenols and tannins correlated best with the percentage increase in gas production on inactivation of tannins by PEG in the *in vitro* gas production method, suggesting that these assays, which are the simplest amongst the group of assays used in the first phase, provide useful information on the biological activity of tannins in the rumen and also in the whole gastrointestinal tract. In one study using leaves from 37 tree and browse species, it was concluded that samples containing total phenols and tannins levels (measured according to the methods outlined in this manual) up to 4% and 2% respectively are not expected to precipitate protein nor cause increases in gas production on the addition of PEG in the *in vitro* gas production method, and, therefore, are not likely to adversely affect ruminant productivity.
- A spectrophotometric method based on rhodanine and an HPLC method for gallotannins have been compared for 38 tree and browse samples. The

specificity and sensitivity of the HPLC method were higher than those of the rhodanine method. However, similar results were obtained by the two methods for the samples containing considerable levels of gallotannins of physiological significance. Using the HPLC method, 24 samples had negligible gallotannin levels (as gallic acid equivalent) (<0.1%), 9 samples between 0.1 and 0.5%, *Dichrostachys cinerea* 0.73%, *Acacia giraffae* and *Calliandra calothyrsus* 2% and 1.6% respectively, *Eucalyptus macrophylla* 3.6% and *A. hockii* 14%.

- A new and simpler spectrophotometric method, based on methanolysis of hydrolysable tannins to methyl gallate and followed by its reaction with potassium iodate, has been developed under the project. These methods should be included in the battery of selected methods for use in the second phase.
- The potential of tannins for use as anthelminths was recognised by the group, but it was not included as a mandatory component in the second phase due to the need to focus efforts and the time constraint.

16.3. Recommendations

- A battery of tannin assays: total phenols, total tannins, condensed tannins, radiolabelled BSA method, and percentage increase in gas production on inactivation of tannins by PEG is required to predict the biological response of feeding tannin-containing diets to ruminants.
- Tree and shrub foliage should be screened for hydrolysable tannins, using any of the three methods (rhodanine, HPLC, methanolysis), to determine the potential toxicity of these feed resources.
- For initial screening of a large number of samples, the two simple assays: total phenols and total tannins could be undertaken.

16.4. Objectives and studies for the on-going second phase

- The main thrust in the second phase is on using the above mentioned validated tannin assays to evaluate strategies to detanninify tannin-rich tree leaves and browses using simple and economically viable approaches, and to exploit the full benefits of tanniniferous plants as animal feed supplements and as strategic feed reserves in situations of fluctuating nutrient supply. All participating groups are working on this common theme and studies are being conducted on high tannin forages of economic importance for the region. Other nutrients and antinutrients are also being considered while studying the effects of tannins *in vivo*.
- Studies will also be conducted on: mechanisms of adaptation to tannins by ruminants; fate of tannin-inactivating agent PEG, PEG-tannin and PEG-tannin-protein complexes in soil using ^{14}C-PEG; simplification of ^{125}I-labelled BSA method; and development of a radiolabelled BSA-based model to investigate the mechanism of tannin interactions in the

gastrointestinal tract and for exploiting various tannins for enhancing protein availability in ruminants.

17. LIST OF PARTICIPANTS IN THE CO-ORDINATED RESEARCH PROJECT

A. Ray Till	Department of Agronomy and Soil Science University of New England Armidale, NSW 2351, Australia
David McNeill	Senior Lecturer in Ruminant Production Faculty of Veterinary Science The University of Sydney Camden, NSW 2570, Australia
M. Raisul Alam	Department of Animal Science Faculty of Animal Husbandry Bangladesh Agricultural University Mymensingh 2202, Bangladesh
Dorinha Miriam.S. Vitti	Centro de Energia Nuclear na Agricultura Secao de Ciencias Animais Universidade de Sao Paulo Av. Centenario 303, Cx. Postal 96 13.400 Piracicaba Sao Paulo, Brazil
Brian McBride	University of Guelph Professor, Ruminant Nutrition Department of Animal & Poultry Science Guelph, Ontario N1G 2W1, Canada
Klaus Becker	University of Hohenheim (480b) Institute for Animal Production in the Tropics and Subtropics Animal Nutrition Department 70593 Stuttgart, Germany
Elizabeth Wina	Research Institute for Animal Production Balai Penelitian Ternak P.O. Box 221 Bogor 16002, Indonesia

Hichem Ben Salem — INRAT, Laboratoire de Productions Animales et Fourragères
Rue Hédi Karray, 2049 Ariana, Tunisia

Ahmet Oncuer — Kafkas University
Faculty of Veterinary Medicine
Kafkas University
36040 Kars, Turkey

Irene Mueller-Harvey — The University of Reading
Department of Agriculture
Earley Gate
P.O. Box 236, Reading RG6 2AT
Berkshire, United Kingdom

Thomas Acamovic — Dept of Biochemistry and Nutrition
SAC
Auchincruive
AYR
KA6 5HW
Scotland, United Kingdom

Timothy Smith — Department of Agriculture
University of Reading,
Early Gate,
P.O. Box 236
Reading Berks, RG6 2AT
Berkshire, United Kingdom

Ann E. Hagerman — Department of Chemistry
Miami University
112 Hughes Hall
Oxford, Ohio 45056, USA

Harinder Makkar — Animal Production & Health Section
Joint FAO/IAEA Division of Nuclear Techniques in Food and Agriculture
P.O. Box 100,
A-1400 Vienna, Austria